Reflective Learning

An essential tool for the self-development of health and safety practitioners

Reflective Learning is the essential reference for health and safety practitioners wanting to develop their professional skills and practice. Whether you are a new practitioner looking to expand your knowledge or an experienced professional seeking to build on existing skills, this book will be indispensable.

Step by step, *Reflective Learning* guides you through the principles to help you to learn and improve your ability to reflect on your past experiences. The use of clear explanations, diagrams and practical tools throughout help you to improve your understanding and advance your professional development.

- ▶ The only book on reflective learning to focus on health and safety
- ▶ Written by experts in the field of health and safety
- ▶ A cost effective way of learning and developing for health and safety practitioners

Teresa Budworth is a Chartered Safety and Health Practitioner, Chartered Fellow of the Institution of Occupational Safety and Health (IOSH) and a Chartered Director. During a 35-year career in health and safety, she has specialised in safety consultancy. Her work on competence, education and training culminated in her appointment as Chief Executive of NEBOSH; the National Examination Board in Occupational Safety and Health, in 2006.

Waddah Shihab Ghanem Al Hashemi is the Chief EHSQ Compliance Officer and Director, EHSQ Compliance for the ENOC Group in Dubai, United Arab Emirates. He is the chairman of various committees within the Organisation, including the Wellness and Social Activities Program which serves over 6000 employees. He is also the Vice-Chairman of the Dubai Centre for Carbon Excellence PJSC "Dubai Carbon". He is a Fellow of the Energy Institute and an Associate Fellow of the IChemE in the UK.

W0234694

Reflective Learning

An essential tool for the self-development of health and safety practitioners

Teresa Budworth
Waddah Shihab Ghanem Al Hashemi

 Routledge
Taylor & Francis Group

LONDON AND NEW YORK

First published 2015
by Routledge
2 Park Square, Milton Park, Abingdon, Oxon OX14 4RN

and by Routledge
711 Third Avenue, New York, NY 10017

Routledge is an imprint of the Taylor & Francis Group, an informa business

British Library Cataloguing-in-Publication Data
A catalogue record for this book is available from the British Library

Library of Congress Cataloging-in-Publication Data
Budworth, Teresa, author.
Reflective learning : an essential tool for the self-development of health and safety practitioners / Teresa Budworth and Waddah Shihab Ghanem Al Hashemi. – First edition.
p. ; cm.
Includes bibliographical references and index.
I. Shihab Ghanem Al Hashemi, Waddah, author. II. Title.
[DNLM: 1. Occupational Health--education. 2. Learning. WA 18]
RM735.3
615.8'515–dc23
2014000695

ISBN: 978-0-415-71551-5 (pbk)
ISBN: 978-1-315-85868-5 (ebk)

Typeset in Univers by
Servis Filmsetting Ltd, Stockport, Cheshire

This book is dedicated to health, safety and environmental practitioners; those already in the profession, and those yet to join, whose vocation protects people, families and society from life-changing injuries and ill health caused by work, and preserves our world for future generations.

Contents

Figures

Tables

About the authors

Teresa Budworth is a Chartered Safety and Health Practitioner, Chartered Fellow of the Institution of Occupational Safety and Health (IOSH) and a Chartered Director. During a 35-year career in health and safety, she has specialised in safety consultancy. Her work on competence, education and training culminated in her appointment as Chief Executive of NEBOSH; the National Examination Board in Occupational Safety and Health, in 2006.

Prior to joining NEBOSH, Teresa combined management of Norwich Union Risk Service's Consultancy operation with her post as a non-executive Director and Trustee of NEBOSH and was Senior Examiner for Diploma Part One from its inception in 1997.

For eight years she was a Visiting Senior Teaching Fellow and member of the Examination Board for postgraduate courses in Occupational Health at the University of Warwick's Medical School. She chairs the Royal Society for the Prevention of Accidents (RoSPA) National Occupational Safety and Health Committee and the judging panel for RoSPA's annual occupational safety and health awards. She has also served as a judge for the SHP IOSH Awards and the International Fertilizer Association Green Leaf Award.

Teresa is Vice-Chairman of the Board of OSHCR Ltd, the body running the Occupational Safety and Health Consultants Register, which was established in 2011 by the UK's Health and Safety Executive and a number of professional membership bodies representing safety and health practitioners. She is also an elected member of IOSH's Council.

Waddah Shihab Ghanem Al Hashemi is
the Chief EHSQ Compliance Officer for the
Emirates National Oil Company (ENOC) Group.
He is the chairman of various committees
within the organisation, including the Wellness
and Social Activities Program which serves over
6000 employees.

Waddah is an Environmental Engineer and
graduated from the University of Wales, Cardiff.
He began his career working as a consultant
for Hyder Consulting Middle East and later
transitioned into the first oil refinery in Dubai,
UAE, ENOC Processing Company LLC (EPCL),
during the construction and pre-commissioning phases. Waddah became the
HSE Supervisor/Coordinator for EPCL and was in charge of all environmental,
health and safety as well as fire protection aspects of the refinery's
operations, maintenance and engineering. While working, he studied
intensively and obtained Diplomas in Environmental Management and Safety
Management from the UK. Waddah was awarded an MSc in Environmental
Science with distinction by the University of the UAE. He has obtained
an Executive MBA through Bradford University in the UK in which he
specialised in organisational safety behaviour. His current doctorate research
work with the Bradford School of Management, Bradford University, UK, is
focused on operational corporate governance and safety leadership.

Waddah has broad experience and working knowledge of environmental
management, safety management and occupational health management
systems. His contribution to the HSE field includes more than 70
presentations and technical papers at various local, regional and international
conferences and forums. Waddah has been a member of the Executive
Committee of the Emirates Safety Group, and serves as an Advisor on the
Higher Colleges of Technology (HCT) Environmental Sciences Program
Committee.

He was a member of the Petroleum and Lube Specification Committee
for the Government of the UAE, and more recently became a member of
the Dubai Demand Side Management and HSE Committee of the Dubai
Supreme Council of Energy (DSCE) and the HSE Committee. Waddah is also
the Vice-Chairman of the Board of Directors of the Dubai Centre for Carbon
Excellence PJSC, a Fellow of the Energy Institute, and an Associate Fellow
of the IChemE.

Preface

Some years ago, we (the authors) started a dialogue on safety practice and how it has developed. We shared the view that Health, Safety and Environmental (HSE) practitioners have become more effective and professional over the years. This was at the ADIPEC 2011 HSE Forum which was entitled "Leadership, Strategies and Innovation to Meet Health, Safety & Risk Management Challenges of a New Era". As expected, many of the attendees were safety practitioners and managers.

Teresa's keynote speech was "View from the Top: Backing Leadership with Critical Actions that Promote, Support and Reinforce a Safe Work Environment." Waddah's presentation focused on a case study of a small-scale pilot to measure safety climate within a chemical plant workforce. It was entitled "Assessment of Safety Culture through Perception Studies – Using Quantitative Methods – Case Study from Management Research in the Emirates National Oil Company (Ltd) LLC Group of Companies". One of the key aims of that presentation was to outline the importance of using scientific methods pragmatically to gather hard data on safety issues.

During the course of the conference we had a long discussion about how much real development was taking place with safety practitioners around the world. The questions we debated included the extent to which academic education, vocational training and attending conferences helped safety practitioners really add value to their organisations; both in terms of making them more effective at influencing an organisation to raise standards and advising on appropriate technical responses to risk. We found ourselves reflecting on our own experiences and careers and our most effective development experiences as safety practitioners.

Our discussion led us to conclude that our personal experiences as safety practitioners were at least equal in importance to our development as our academic and vocational training. But how did we learn from these experiences, and did we actually learn as much as we could? Did we also, at the time of having the experience, realise that in the future, that very experience would lead to our eventual refinement as safety practitioner and manager? This discussion led us to develop a paper on reflective learning which was presented at the February 2012 American Society of Safety Engineers, Middle East Chapter Conference in Bahrain, and published in the Proceedings.

Teresa delivered the paper and the feedback from the audience was very encouraging. We had presented something new to many HSE practitioners. We felt that the value of a reflective approach to learning was something worth sharing with more HSE practitioners so we decided to go deeper.

Later that year we prepared an outline of a book and presented it to the publisher, who sought some independent reviews of our proposal. There was great feedback from the independent reviewers which inspired us to continue. The outcome is this handbook which we hope will be a practical tool for self-development and motivate HSE practitioners to engage with the practice of reflective and experiential learning.

While some theory is discussed in the earlier chapters of the book, we have tried as much as possible to illustrate the academic principles with practical examples drawn from both our own experience and from the experiences of a number of other HSE practitioners with whom we conducted semi-structured interviews. We hope that the examples we give and the case study discussions show clearly the uses of reflective learning. However, we feel the greatest value of the book comes in the form of the reflective tools that we present. Not all of the tools will suit every practitioner, but we hope they will inspire readers to try out new approaches for themselves.

Acknowledgements

We would like to thank the safety, health and environmental practitioners and related professionals who helped us with our journey as HSE practitioners and/or with the preparation of this book by sharing their reflections, experiences, ideas and inspiration. Our particular appreciation goes to Stephen Asbury, Andrew Ashford, Bashyr Aziz, Roger Bibbings, Richard Booth, Tim Briggs, Neil Budworth, Alia Mubarak Busamra, Ray Faulkner, Roy Featherstone, Ray Hurst, Tony Ireland, Anabel Jay, John Kersey, Richard Lovering, Patrick McLoughlin, Muhammed Noh, Charlie Pay, Jim Pearce, Ron Preston, Yasser Rahim, Kanak Rao, George Robertson, Mike Stevens, Nikiforus Stouraitis, Carlos Tan, Lawrence Waterman and Barry Wilkes.

We would also like to thank Sadé Lee for her patient transcription of the interviews.

Lastly, but most important of all, our gratitude goes to our families who supported us throughout this process.

CHAPTER 1

Purpose and value of this book

Over the last two decades, health, safety and environmental management have received greater recognition as professional disciplines in their own right. Health, Safety and Environmental (HSE)[1] practitioners can choose to qualify through a variety of professional routes in many countries around the world, and Board Certified or Chartered Practitioner status is available to them.

Organisations employing HSE practitioners have also begun to place a great deal more value on the set of skills that they bring to the job, as they have increasingly begun to appreciate the critical nature of what is involved. HSE practitioners now operate at more senior levels within organisations, and career progression to board level is a real possibility.

In HSE, as in many professions, the concept of continuing professional development post-qualification has also come to the fore over the last 20

---

[1] We have chosen to refer to Health Safety and Environmental (HSE) practitioners throughout this book. We recognise that they may be considered as three separate professional disciplines, and individuals may practise in only one or two of these disciplines. Some practitioners may combine responsibility in one or more of these disciplines with quality or security management. Other practitioners may have responsibility for HSE, perhaps in a very specialist niche area as part of another professional role, perhaps as an engineer or occupational psychologist. We hope that the principles we discuss will be equally applicable to all professionals who make their contribution in this important field.

or so years, and for the HSE practitioner this can include management skills as well as developing their technical knowledge to keep pace with the changing world of work, technology and emerging risks.

Reflective or experiential learning has been an important development in the further education of a number of professions. Progressing from the academic research carried out on the topic in the early 1980s, it has become an integral part of professional training in medicine, nursing, teaching and management. Many of the traditional professions have a well-defined training route, including acquisition of the underpinning academic knowledge combined with supervised practice to develop the necessary skills to operate as an independent professional. Reflective learning forms a key part of this initial professional development, and the skills in becoming a "reflective practitioner" then endure throughout a career and are applied in continuous professional development.

Learning from your on-going professional practice is just as important as the initial technical training. However, for those working in HSE as practitioners and managers, becoming a "reflective practitioner" may not have formed part of their initial development. Hence very few HSE practitioners recognise the significance and power of using reflective learning tools, which denies them the opportunity to maximise the significant development opportunities from their own practice which would in turn end up making them better HSE practitioners.

As with practitioners in any field, recruiting employers value experience in managing and handling various situations and issues. However, while many practitioners may have gone through similar experiences, in similar settings in relatively the same period of time, we very often find that the experience has shaped people differently. But the question is, has the experience actually shaped you, or was it how you reflected on that experience and gained from it that has ultimately made that improvement and brought about that richer understanding? Often it is that experience and understanding that allow us to manage the next situation in a better or at least a more effective way. Much of what we are able to do, analyse or perform is linked to our comprehension, intelligence and knowledge. However, sometimes effectiveness and efficiency are achieved by a "prior" knowledge which was not obtained through either academic or vocational education. It was often obtained through a past experience, on which we reflected and then drew conclusions from. This is a natural process and often not a totally conscious one. However, the process of reflection can be made into a conscious process where it becomes part of the self-development of the person.

This book covers the theory underpinning reflective practice, the different tools which are available to the practitioner and provides an explanation of why reflective learning is probably one of the single most important learning resources and tools that HSE practitioners can use to improve their self-development and knowledge.

The book draws on examples and case studies from a number of HSE practitioners illustrating how reflective learning is being practised. These were gathered in a series of semi-structured interviews as well as from both the authors' personal experiences and practice over the past three decades.

Structure of the book

The book is divided into seven chapters. Chapter 2 is an introduction to this book and explains why the authors felt it was important to develop such a guide for practitioners. It also explains the methodology used to develop this book, and in particular the examples and case studies in the later chapters to illustrate the importance and significance of reflective learning.

In Chapter 3 we start to describe in greater detail the theoretical and research-based foundations of reflective learning. This chapter seeks to explain how reflective or experiential learning is connected to more general theories of learning.

Chapter 4 discusses the different types of learning available to the HSE practitioner, and how practitioners can use these methods in their own development. These are discussed in the context of both initial development and maturing as a seasoned practitioner either as a specialist or a generalist and also later on as managers and directors.

Then, in Chapter 5, the importance of reflective learning for HSE practitioners is discussed at length with our insights from the interviews that have taken place with various practitioners. The chapter addresses the development of practitioners in a technical and managerial capacity and also the impact of reflective learning on developing behavioural skills. The advantages and benefits of reflective learning for practitioners as a tool of self-development are naturally emphasised.

In Chapter 6, we address the different types of reflective learning and how they can be employed, including using self-development tools; working with a mentor; reflecting in groups; and also something very appropriate for HSE practitioners which is reflective practice in projects.

The final chapter draws on our key conclusions from much of the theory, writings and, most importantly, our personal experience and the semi-structured interviews undertaken in the course of writing this book.

We believe that this book will be a valuable resource to HSE practitioners, and those working with practitioners on their development. While reflection is in no way a new or novel concept, using reflection as a learning tool for development and self-improvement in a conscious and systematic way can help improve the effectiveness of the HSE practitioner for their own benefit as a professional building a career and for the benefit of those whom they protect in exercising their professional skills.

CHAPTER 2

Introduction to reflective learning

When I graduated in the early eighties with a degree in health and safety, it took me ages to get my first proper job. At that time it was not a graduate level role. In fact, I actually saw a job advertised for "Car park attendant and safety officer". Needless to say, I didn't apply for that one.

Health and Safety Consultancy Manager

A significant change in the world of health and safety over the past 40 years has been the elevation of the discipline to a profession in its own right. In the United Kingdom, in 2005, professional members of the Institution of Occupational Safety and Health (IOSH) were for the first time awarded the status of Chartered Members in recognition of this fact.

It is worth considering what it means to be a "professional" or a member of a profession. The *Oxford English Dictionary* defines "professional" as an adjective as:

Relating to or belonging to a profession . . . worthy of or appropriate to a professional person; competent, skilful, or assured.

And as a noun:

A person engaged or qualified in a profession . . . a person competent or skilled in a particular activity.

Everett Hughes[1] explains: "A profession delivers esoteric services – advice or action or both – to individuals, organisations or government; to whole classes or groups of people or to the public at large." He elaborates: "Even when manual, the action – it is assumed or claimed – is determined by esoteric knowledge, systematically formulated and applied to the problem of the client."

His view was that professional bodies and the professionals belonging to them have a kind of bargain with society, in that, in exchange for their skills and extraordinary knowledge which they exercise for the good of society, they are allowed a high degree of autonomy to self-regulate, to determine who is allowed to enter the profession and under what circumstances. This gives rise to familiar structures in professional bodies of prescribed qualifications, peer evaluation of entry and codes of behaviour or ethics.

Edgar Schein[2] contends that there are three components to professional knowledge:

1 An underlying discipline or basic science component upon which the practice rests or from which it is developed.
2 An applied science or "engineering" component from which many of the day-to-day diagnostic procedures and problem-solutions are derived.
3 A skills and attitudinal component that concerns the actual performance of services to the client, using the underlying basic and applied knowledge.

Schein's view was that these components form a hierarchy. The application of basic science yields engineering, which in turn provides models, rules and techniques used in everyday practice. The actual performance of services "rests on" applied science, which in turn rests on the foundation of basic science. Basic science is therefore at the top of the hierarchy. The usual curriculum of professional education follows this hierarchy. First, students are taught the underpinning basic science, then the relevant applied science, and finally they are exposed to a practicum in which they are presumed to learn to apply classroom knowledge to the problems of practice. In this traditional model there is not a great deal of interaction between the practitioners of research and the practitioners in the field.

Influential work on the education of professionals by Donald Schön[3] suggests that this is a flawed conception of professional competence. In the real world of practice, professional competence is not just about the applications of theories and techniques arising from scientific research. In their formal training, HSE practitioners learn that safety issues can be resolved by identification of hazards (naming and framing the problem), evaluation of the risk involved and the application of a solution based on a hierarchy of control measures, which is commensurate with the risk and deals with the root cause of the problem. This is "technical rationality" applied to health and safety at work.

In practice, many problems do not neatly fit this model and may involve situations which are, in Schön's words, "unique, uncertain, conflicted". However, the ability to resolve such issues, and to embed practices within an organisation which can prevent the recurrence or give a framework to deal with them, are central to competent practice as a safety professional. They require the ability to develop and test new forms of understanding and action where familiar categories and ways of thinking fail.

In practical educational terms, the skills to deal with more complex problems are hard to teach. The skills involved are difficult to define, crystallise and incorporate into a formal course of training.

Logically one would imagine that by studying the activity of competent practitioners, it would be possible to distil the skills they apply, or to ask them to describe their approaches. In practice, this is more difficult. Many of the skills are based on "knowing in action". That is, in exercising such skills, it is difficult to articulate exactly what it is that is being done. They are executed as a smooth series of actions without any conscious control. Examples of such "knowing in action" are riding a bicycle, operating the gears on a car or even walking. It just "feels right".

However, the extension of such skills also comes about by "reflection in action". The "knowing in action" works for normal day-to-day activities, but then if a problem arises which is outside of the normal situation, this requires a different approach. This can sometimes be an unexpected event. For the safety practitioner, it may be an accident, a case of ill health or a "near miss" which is unusual or outside of their previous experience. Determining the causes and designing means of preventing a recurrence can require a great deal of intellectual effort.

Jens Rasmussen[4] describes three levels of human performance, categorising them as *skill-based, rule-based* and *knowledge-based*. Rasmussen's model of human performance is presented in Figure 2.1. The three levels can be distinguished by the type of situation; whether

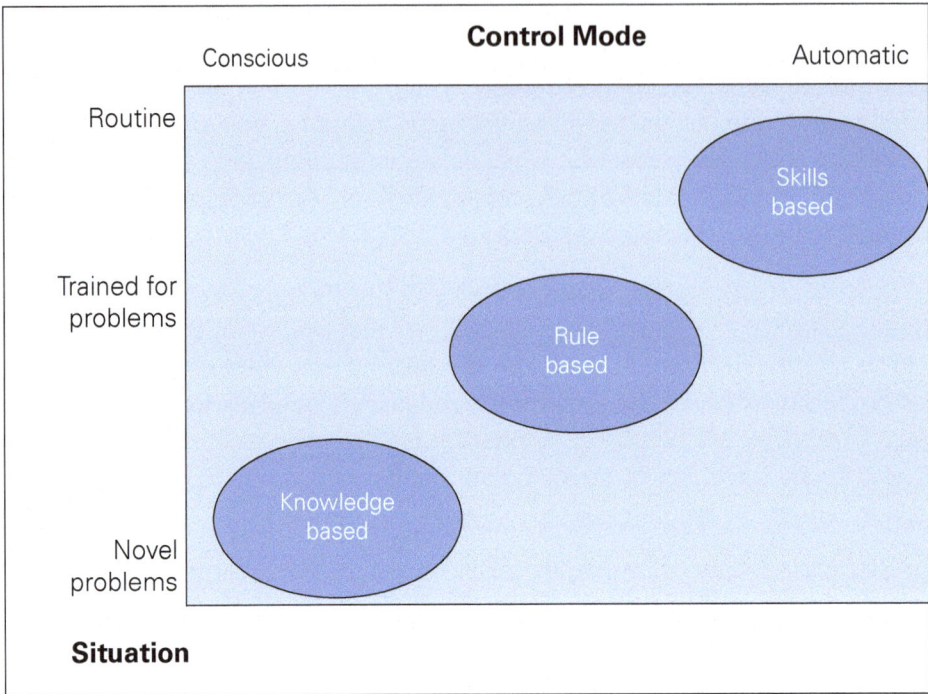

Figure 2.1 Rasmussen's model of human performance
Source: Adapted from Rasmussen (1982) and Reason (1990).

it is routine or a novel one; and the degree of conscious control we are exercising, in particular whether this control is conscious or automatic.

When we are in conscious control mode, we are paying a great deal of attention to what we are doing. However, attention needs to be considered a limited resource: If we are devoting all of our attention to one thing, such as solving a novel problem in a knowledge-based operation, we cannot direct our attention to other things. The automatic mode is the opposite in many respects. We are carrying out routine actions that we are very skilled at, and do not need to pay conscious attention to doing them, we do them automatically. Hence we can carry out a number of such activities in parallel, as in the old expression, "You can chew gum and walk at the same time." This is the "knowing in action" as described by Schön (1987).

Examples of these different levels of operation can be experienced when driving a car. When we are learning to drive, we find ourselves operating the controls in a very slow and halting way, hesitating, as we work out what the effect of our actions will be on the motion of the vehicle. We take our time consciously driving, as we are in knowledge-based operation. As we become skilled as drivers, we no longer have to consciously think about how we operate the controls of the car. Actions like steering, accelerating and braking become automatic. We form the intention, but the process of moving the steering wheel is below our conscious level of control. We are in skills-based operation. This leaves our attention available for those things that we do have to pay attention to when driving, like other road users and traffic.

Rules-based operation is situated between the other two modes. It occurs when we need to change our actions because of a change in circumstances. It is likely to be a situation or problem that we have encountered before, and we apply a known rule to it. If *this* is the situation, then we will do *that*. For example, when driving, if we wish to turn on to a main road, we monitor the flow of on-coming traffic. Our rule is that if there *is* on-coming traffic, then we wait. If there *is not* on-coming traffic, then we complete our manoeuvre on to the main road.

As we become more expert, our repertoire of useful rules increases, as we encounter situations which are inherently similar to those we have encountered before.

Many safety practitioners will recognise Rasmussen's model and its use by James Reason in his work on types and causes of human error.[5] Knowledge-based operation is prone to errors – classified by Reason as "mistakes", which can arise from, for example having insufficient underpinning knowledge or insufficient time to find an appropriate answer. We also have a marked preference for applying existing rules even if they may not be appropriate to a changed situation.

In fact, there is an old saying that runs, "He who never makes mistakes, never makes anything", which refers to what is called "trial and error" learning. The starting point may be a novel situation, such as an unusual accident as described above. The problem or situation or new information can be reflected upon and methods considered of overcoming the problem, which can then be tested to see if they do in fact solve the problem. This may be a continuous process and is well described by Honey and Mumford[6] in their development of Kolb's learning cycle (Figure 2.2). The stages of Kolb's learning cycle[7] are: (1) concrete experience, which is doing/having an experience; (2) reflective observation, which is reviewing/reflecting on the experience; (3) abstract conceptualisation, which is concluding/learning from the experience; and, finally, (4) active experimentation, which is trying out what you have learned, which in turn becomes the next concrete experience, and the learning cycle starts again.

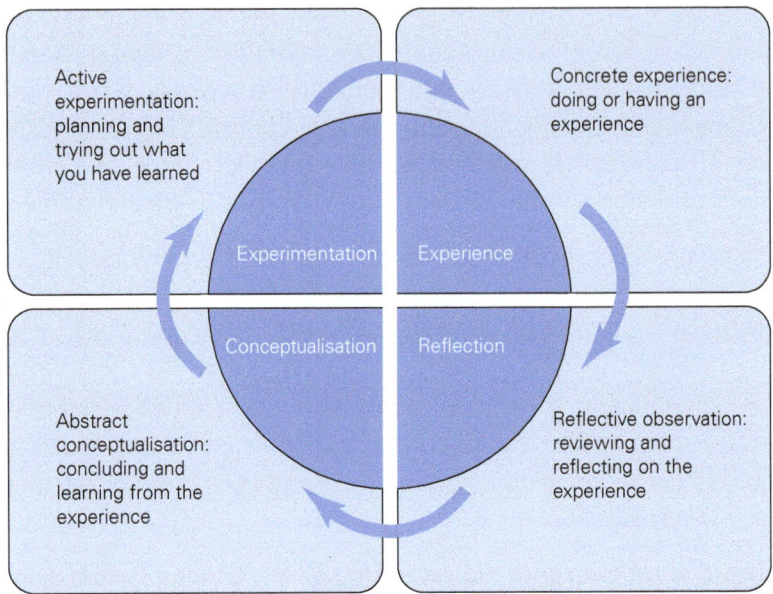

Figure 2.2 Kolb's learning cycle
Source: Kolb (1984).

In this model, skills are not "taught" but are learned, and learning takes place by doing. These concepts underpin what is referred to as *experiential* or *reflective* learning. This can help the HSE practitioner to develop a repertoire of approaches as his or her experience and skill build over time. This is not to say that the learning is completely independent. However, the learning is more centred on the learner, sometimes with a coach, a mentor or a guide (see Chapter 6 for more on working with a mentor).

Reflective learning has entered the professional development of a great many disciplines. In management schools, the "reflective paper" may be part of the formal assessment. Nursing practitioners use reflective accounts as part of their development, considering "What have I learned from this? How will it influence my future clinical practice?"

Methodology in developing the book

In preparing this book we considered the traditional methods of educating safety practitioners. We reviewed the literature on reflective learning, and in particular how other professions use reflective learning.

We also conducted semi-structured interviews with ten safety and health and environmental practitioners, to let them describe how they determine their own learning experiences, how they have developed within their profession and the use they make of reflection in their professional development. We have used extracts from the transcripts of those interviews throughout the book where they illustrate the concepts being discussed.

In brief

▶ Traditionally, a professional, in any field, delivers a service – advice or action or both – based on esoteric knowledge, systematically formulated and applied to solving the problem of the client.

▶ In practice, many problems do not neatly fit this model and may involve situations which are "unique, uncertain, conflicted".

▶ The ability to develop and test new forms of understanding and action where familiar categories and ways of thinking fail is central to competent practice as a safety professional.

▶ Developing your professional practice requires "reflection in action", building on your existing understanding, by trying different approaches.

▶ This process, which can be referred to as reflective or experiential learning can be described by the Kolb learning cycle.

Notes

1.   Hughes, E.C. "The Professions", *Daedalus*, 92(4) (Fall, 1963): 655–668.
2.   Schein, E. *Professional Education* (New York: McGraw-Hill, 1974), pp. 43–44.
3.   Schön, D.A. *Educating the Reflective Practitioner* (San Francisco: Jossey-Bass, 1987).
4.   Rasmussen, J. "Human Errors: A Taxonomy for Describing Human Malfunction in Industrial Installations", *Journal of Occupational Accidents*, 4 (1982): 311–333.
5.   Reason, J. *Human Error* (Cambridge: Cambridge University Press, 1990).
6.   Honey, P. and Mumford, A. *Manual of Learning Styles* (London: Peter Honey Publications, 1982).
7.   Kolb, D. *Experiential Learning: Experience as the Source of Learning and Development* (Englewood Cliffs, NJ: Prentice Hall, 1984).

CHAPTER 3

The foundations of reflective learning

Learning has been defined as "a process that brings together cognitive, emotional, and environmental influences and experiences for acquiring, enhancing, or making changes in one's knowledge, skills, values, and world views".[1,2]

Often learning objectives are set out in terms of the following elements that the learner is meant to acquire:

▶ knowledge
▶ skills
▶ attitudes.

This simplistic model corresponds to Bloom's taxonomy (taxonomy means classification). In his (1956) taxonomy, Bloom proposed three domains or categories of learning:[3]

▶ cognitive
▶ psychomotor
▶ affective.

These domains are not mutually exclusive. For example, when learning to drive, the trainee will have to learn the laws related to driving, such as on which side of the road to drive and who has "right of way" at junctions (the cognitive domain), but the learner driver also has to learn how to steer, to

Table 3.1 The cognitive domain of Bloom's taxonomy

Knowledge	Recall information
Comprehension	Understand meaning
Application	Use or apply knowledge in a practical situation
Analysis	Analyse information, relationships and find and apply evidence to support conclusions. "Open-ended" problem-solving
Synthesis	Develop new ideas, models and approaches based on knowledge
Evaluation	Assess effectiveness of concepts, critical thinking, strategic review; judgement relating to external criteria

Source: Bloom (1956).

indicate and to adjust the seat position (the psychomotor domain). It is also desirable that the learner driver comes to appreciate the value of showing courtesy to other road users and even to enjoy driving for its own sake (the affective domain).

Within each domain, there is a hierarchy of learning objectives, each of which needs to be mastered before progress can be made to the next stage of higher objectives. For example, Table 3.1 shows the hierarchy found in the cognitive domain, and these are set out in order from the lowest to the highest.

Often in the education of health, safety and environmental (HSE) practitioners and other professionals, the focus is on acquiring, understanding and applying the underpinning knowledge (the cognitive domain). However, to be an effective health, safety and environmental practitioner, particularly if functioning at a senior level in an organisation, there is a clear need to have higher-order cognitive domain learning. For example, synthesis – to be able to devise new safety management systems, to assess, evaluate and control emerging risks, and evaluation, to critically appraise performance against external criteria and modify the approach of the organisation.

In addition, a range of other skills need to be developed. Implementing change in an organisation requires a combination of cognitive ability and the skills of communication, persuasion and influencing people. It is clear that the learning and development needs of the HSE practitioner cannot be met by one single injection of training and education at the beginning of a career; in other words, development needs to be continuous.

HSE, as a relatively new profession, does not have a clearly defined professional entry route, with distinct, clear career pathways through it.

It may often be the responsibility of the individual practitioner to identify their own learning needs and determine how best to meet them. Reflective learning techniques can assist with this, but the practitioner needs to approach this with an open mind.

In this chapter we aim to review some of the theories of learning which may assist practitioners in thinking about how to develop their reflective skills. In later chapters we will discuss the practical application of these theories into tools and techniques.

There are significant differences between different types of learning. Academic learning is primarily driven by studying and analysis of information and data. It is a very important process of development and establishes foundations that help the learner to build understanding. Higher levels of academic learning extend the boundaries of knowledge and understanding, benefiting not just the learner, but contributing significantly to the continual development of society as well as the academic community.

Experiential or reflective learning can be defined as the "practice of experience-based learning" but it essentially must follow these criteria:[4]

- ▶ the learning that results is personally significant or meaningful to the learner;
- ▶ the learner becomes personally engaged with the learning;
- ▶ a reflective process is involved;
- ▶ there is recognition of the past experiences of the learner and others.

Experiential or reflective learning can build upon the knowledge gained through academic learning, which, as we have seen in Bloom's *Taxonomy of Educational Objectives*, does not just describe being able to recall information, but also being able to take that information and apply it, use it to solve new problems and develop new approaches. Experiential or reflective learning has been described as "[the] knowledge that results from actual observation or from what one has undergone".[5]

Adult learning through experience is very empowering as it opens to the learner knowledge which is familiar, which gives a sense of ownership. Adult learners often require or demand that the application of ideas is tested against their own personal experiences.[6] This makes learning from experience arguably more effective for adults than theoretical/conceptual learning. This is an important consideration in the education of HSE practitioners, most of whom come to the HSE field as a second or third career.

There is a great deal of literature from the early 1980s regarding reflective learning and also a great deal of agreement that reflection on your own practice is important for the development of professionals.[7] The work of Donald Schön[8] in the early and mid-1980s has greatly influenced professional education in the UK.[9] His concept of "the reflective practitioner" has become well established in the fields of teacher training, health care and even finance.

Reflective practice is increasingly significant in developing managers in the context of organisational learning.[10,11] The ability to reflect and develop more effective decision-making processes is what many organisations are looking for in modern managers. This is further compounded in recent times with the fact that in a world of globalisation with rapid developments in the technology of communications and information, there are higher expectations of faster decisions and responses.[12] Furthermore, it has been argued by Barry Smith[13] that it is managers who are able to direct their acts thoughtfully and thus in a calculated way that become more effective at controlling situations, as they start to consciously choose the most effective behaviour when addressing the situation that confronts them. This is a fairly key skill for HSE practitioners; in particular, as they move into more senior positions and move between different organisational cultures, where they need to use influencing skills to achieve results.

Learning describes a process by which practitioners acquire knowledge and skills to enhance overall competence and effectiveness. Experience is an important learning process in which the practitioner engages in a situation and – it is hoped – becomes more proficient at dealing with that situation in the future. However, in the absence of reflection which is the process of standing back from an experience and taking some time to analyse and carefully review it, some of the learning opportunities may be lost. In this analysis, the practitioner draws meaning from the reflection which results in learning.[14]

Reflection does not have to just take place after an experience. It can also occur while the experience is happening. Reflection has been described by Christopher Johns as:

> *being mindful of one self, either within or after an experience, as if a window through which the practitioner can view and focus self within the context of a particular experience, in order to confront, understand and move towards resolving contradiction between one's vision and actual practice.*[15]

In this description, reflection is an introspective process through which the practitioner gains insight which leads to developing practical knowledge and wisdom when dealing with future situations.

A further theory is that there are five stages (or layers) in the reflective learning process:[16]

1 *Reflection-on-action* which is reflecting *after* an event or critical incident.
2 *Reflection-in-action* which is *pausing in* a situation in order to make sense of it, and allow yourself to work towards your desired outcome.
3 As an *internal supervisor*, which means having an internal dialogue with yourself while discussing the situation with someone else.
4 *Reflection within-the-moment* which is being aware of your thoughts, feelings and behaviours as events are unfolding and dialoguing with yourself to ensure you are interpreting and responding appropriately in the situation.
5 *Mindful practice*, which is being aware of yourself in the moment to allow you to work to your best abilities.

Reflection gives the individual choices from a repertoire of approaches. This is particularly important in a rapidly changing environment or one with a high degree of uncertainty.[17] Therefore, a fundamental aspect of management education is the prioritisation of reflective ability as well as embracing change positively in the context of it being a learning opportunity.

The process of reflection can also change not just the choices that an individual can make in a situation, but also their perception of the situation itself. An early definition of reflection which encapsulates this is:

> *the process of internally examining and exploring an issue of concern triggered by an experience, which creates and clarifies meaning in terms of self and which results in a changed conceptual perspective.*[18]

There are three stages in the process of reflection described by this definition which can be elaborated as:

1 *Awareness* – the individual becomes conscious of a particular experience.
2 *Critical analysis of the experience*, which is likely to have been discomforting. This stage includes the critical thinking in which the individual starts challenging their underlying assumptions of their existing knowledge.

3 *Perspective transformation*, which occurs as the individual adopts a new perception of, and thus eventually behaviour towards, a similar situation in the future.

In critical reflection the change occurs as a person starts to challenge their own firmly held beliefs and thinking. This concept is further described in Argyris and Schön's model of single, double and triple loop learning[19] or first-, second- and third-order learning.

First-order learning occurs when some incremental change is made, resulting in an improvement in the way you actually do something. In second-order learning, the change is of a greater magnitude as it effects a change in the theory-in-use as suggested by changes in strategies and assumptions.[20] This involves an individual looking at change within a particular context and leads to a more significant change compared to first-order learning. In third-order learning, the change is in the context itself and this means a very serious and deep-rooted change, in which the individual learning actually embraces new values in the theory-in-use and the learning processes involved.[21] These processes are illustrated in Figure 3.1 and Figure 3.2.

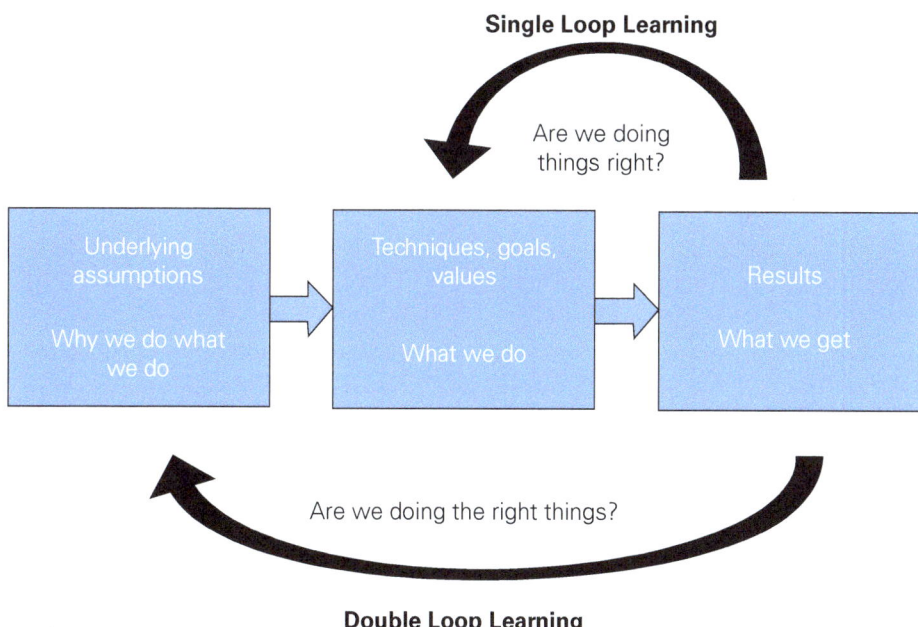

Figure 3.1 Single and double loop learning

Source: Argyris and Schön (1996).

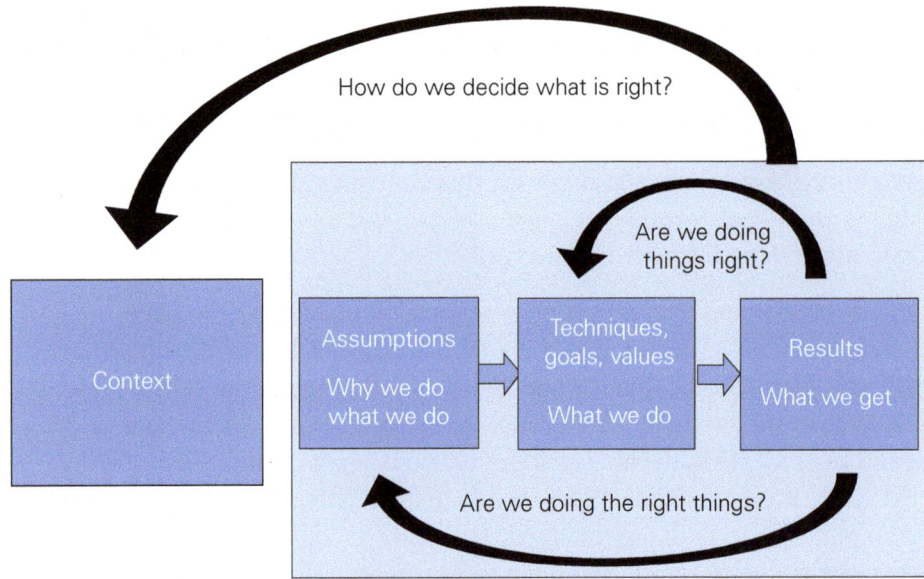

Figure 3.2 Triple loop learning
Source: Argyris and Schön (1996).

We have previously looked at Kolb's learning cycle, which proposes that learning is a cyclical process. This can also be considered a problem-solving process. A person has an experience, and they may be able to identify some key aspects of that experience and compare them with some set of values or their existing interpretation of reality. This leads to identification of the differences and evaluation and identification of the problems (this is the problem finding stage). This focus leads to an exploration of alternative solutions; and a logical assessment of the consequences of alternative courses of action before eventually choosing a solution to execute (this is the problem-solving stage). These stages may not be distinct and can take different durations. With new experiences (including implementing the solution to the problem), the cycle reoccurs. This is illustrated in Figure 3.3, which is based on four stages: (1) acting; (2) valuing; (3) thinking; and (4) deciding, which moves the individual back to the beginning of the cycle. Figure 3.3 is adapted from Bostrom *et al.* (1990)[22] and Scott (2002).[23]

One important element in the model shown in Figure 3.3 is the label applied to the person's role at each stage of the cycle: Activist, Reflector, Theorist and Pragmatist. In Stage 1, the Activist is doing something – having an experience (acting). In the second stage, Reflectors are taking time to think about what happened – reviewing the experience, followed by assimilation

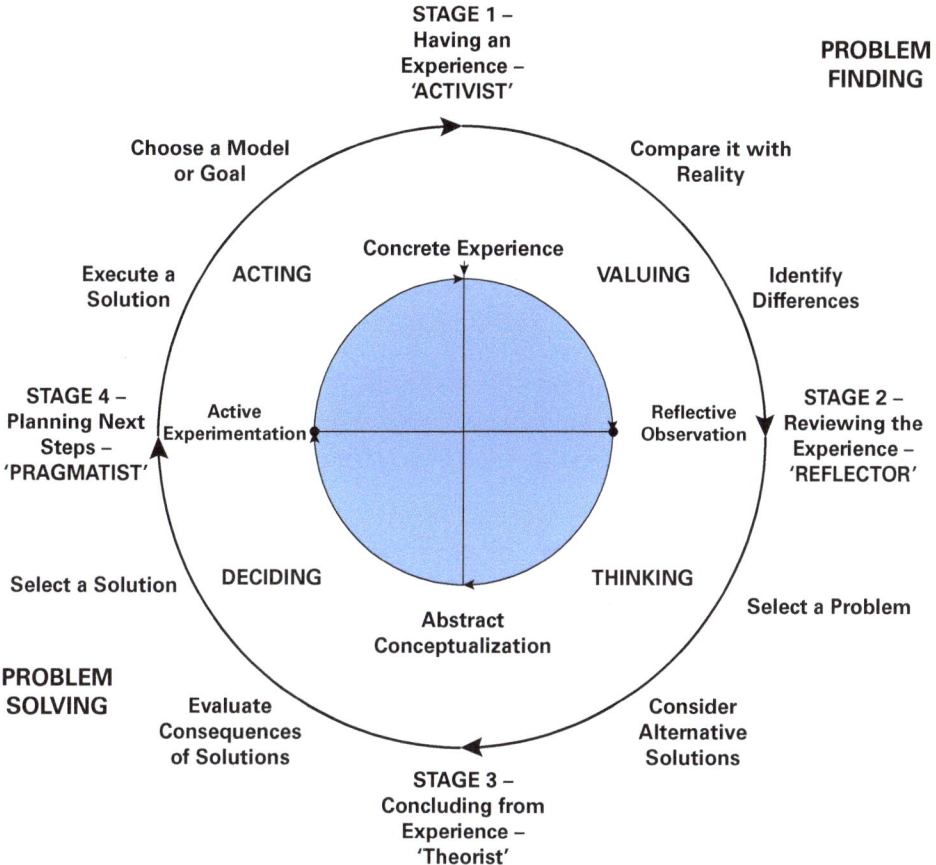

Figure 3.3 Comparison of the experiential learning model and a problem-solving process

Source: Adapted from Bostrom *et al.* (1990) and Scott (2002).

of the experience and integration or fit with existing knowledge (valuing). In the third stage of decision-making, prior to finally selecting the solution, abstract conceptualisation is undertaken and is the persuasive paradigm of the Theorist who ultimately converges to a solution (thinking). Pragmatists are interested in the application of the solution and move to active experimentation (deciding).

For effective learning to take place, the individual needs to move through the cycle and spend time in each role. However, individuals may show a marked preference for one or more stages of the cycle. These have been described as learning preferences or learning styles.[24,25] If there is one learning style that the individual particularly dislikes, then this can actually hinder learning and problem solving. For example, failing to reflect can lead a person to fail to identify all the causes of a problem. Failing to spend time on conceptualising or drawing conclusions can mean that some options to solve a problem are not identified or explored and may be prematurely rejected. Being uncomfortable as a pragmatist may mean testing out how proposed solutions may work in practice is omitted.

Understanding your learning style can help you to select learning activities which best fit your preferences. This is explored further in Chapter 5.

Understanding the areas to which you are not naturally inclined can help you to become a better reflective practitioner. You may develop your learning abilities by consciously choosing to spend time, for example, reflecting, or looking for links with your existing knowledge or considering alternative solutions or looking at the practicalities of your proposed remedy.

There has been a great deal of research on methods or tools[26] to assist learners to develop their reflective or experiential learning ability. The tools for reflective learning seek to help the individual to spend time at each stage of the experiential or reflective learning cycle, encourage deeper understanding and may bring about transformation. The tools employed help learning by making the process more efficient and focusing on *critical* reflection, which incorporates questioning of assumptions and consideration of other perspectives. As we saw when considering single, double and triple loop learning, unless we challenge our own "taken-for-granted" assumptions,[27] we cannot be receptive to other ways of reasoning and behaving.

The practical application of reflective learning tools by the HSE practitioner will be discussed in greater detail in the later chapters, but it is useful at this point to look at some of the main techniques advocated by researchers in the fields of learning, management development and professional education. The following techniques have been found to be successful by many practitioners, and are more fully discussed in Chapter 6.

- ▶ *Story-telling* is where the individual narrates events or articulates the problem by telling a story. The story may be documented in a learning journal or "told" to a coach. The meaning, values, norms, contradictions, and alternative points of view can be developed within the story or explored with a coach or mentor.
- ▶ *Reflective conversations* are generally done in pairs with a facilitator who listens and questions, discusses and advises. These involve exploring contradictions, doubts, dilemmas, and possibilities in order to think more critically about one's values and the effect of one's actions on others.

▶ *Reflective dialogue* takes place in groups, but here, the participants take it in turns to speak and the facilitator encourages the group to challenge taken-for-granted assumptions.

▶ *Reflective metaphor* is where the learner is encouraged to think of a metaphor to describe themselves, the situation or event. This then triggers analysis or discussion. It can be used in story-telling, or in a reflexive conversation or reflective dialogue.

▶ *Reflective journals* may be a private tool, used only by the individual, or some or all of the contents may be shared with a mentor or coach. The journal is used to document experiences and the learner's reflections on those experiences.

▶ *Critical incident analysis* is the application of reflection to an event or incident which may be a turning point or a catalyst for personal or organisational change. The event may be discussed with a coach, or form the topic for reflective conversations or dialogue, or be captured in a story or within a reflective journal.

▶ *Repertory grids* are based on work by Kelly[28] on the psychology of personal constructs. This is a tool for identifying the basic assumptions or constructs that help you to make sense of the world in order to anticipate future events. Constructs may be considered as being dichotomised or bipolar, and in making a grid you plot where you are on that construct's continuum. The construct is discussed with a facilitator or group to identify norms, values and beliefs. Critical reflection can lead to a change in your view of the world, and to the use of a more useful construct. This is a technique sometimes used in some schools of psychotherapy to challenge the assumptions that a distressed person is using which are unhelpful to them.

▶ *Concept mapping* requires more conceptual thinking. The learner maps a series of concepts hierarchically and links the concepts with arrows labelled with explanatory phrases emphasizing the "how" and "why" of links. It may promote reflection by the individual or in groups by setting out explicitly the issues, values and assumptions found in a situation and provide a focal point for reflection. However, not everyone is comfortable with this approach which can limit its usefulness.

Reflective journals are probably the most commonly used tool to enhance reflective practice.[29] They should and may be distinguished from diaries and logs by their purpose, which is essentially as a tool for reflection in which entries are made over a period of time. They contain reflections on experiences with the intention of enhancing personal learning.

Writing in general terms about experiences/thoughts can be a very liberating experience and can enhance learning. Journal writing enhances learning through dedicating and demanding time and intellectual space. The journal can be a place not just for recording events, but a place where the experience is processed and made sense of, employing critical thinking.

Writing in itself as an activity forces time for reflection; it forces some order of thoughts; to make the recording of an experience as an account more understandable. Writing also allows the recording of a train of thoughts, the capture of different ideas and can be a creative activity which trains the mind and enhances reflection through the slowing down of the pace in thinking.[30]

There are strong arguments for keeping journals private, because it means that writing is not as constrained as it would be if it were revealed to others. For example, it is appropriate to explore in a learning journal negative feelings about your own performance or the role others may play in a situation. The nature of the reflection may be to consider alternative ways of handling the problem and working with difficult individuals. If that individual is making a formal evaluation about your capabilities at work, sharing your reflections may not enhance your career. On the other hand, sharing your journal entries with a coach or mentor, who is not formally assessing you, can help draw meaning from your entries.

Finally, reflective learning can apply not just to the individual but also to the organisation or groups within it. One definition of a "learning organisation" is: "an organisation skilled at creating, acquiring and transferring knowledge, and at modifying its behaviour to reflect new knowledge and insights".[31]

Formally reviewing the "lessons learned" should be a key part of all projects.[32] Critical reflection and reflective learning are vital in the context of continual improvement for project managers, and this includes HSE practitioners who are working on delivering any change in their organisation's approach to managing safety, health or the environment. In many cases, reflection is impeded by a busy schedule and the fast-changing nature of projects, with organisational rewards for getting the job done on time rather than on thinking too long about it. However, it is imperative that reflection takes place both regularly during and after a project and this can be scheduled in as a task within the project plan, to allow post-project analysis and closing of the learning loop.

For practitioners to develop, they must take the time to reflect and in order to enhance their learning from experiences. Without meaningful reflection, managers and practitioners will eventually start to lose competence with time.[33]

In brief

- ▶ Reflective learning can be defined as the "practice of experience-based learning" that essentially has the following features:
  - ▷ the learning that results are personally significant or meaningful to the learner;
  - ▷ the learner becomes personally engaged with the learning;
  - ▷ a reflective process is involved;
  - ▷ there is recognition of the past experiences of the learner and others.
- ▶ In the absence of reflection which is a process of standing back from an experience and taking some time to analyse and carefully review it, some of the learning opportunities may be lost.
- ▶ The process of reflection can change not just the choices that an individual can make in a situation, but also their perception of the situation itself. In critical reflection the change occurs as a person starts to challenge their own firmly held beliefs and thinking.
- ▶ Reflective learning and problem-solving processes are closely linked and a greater understanding of the stages of learning and awareness by the practitioner of his/her own learning style preferences can be very valuable for self-development.
- ▶ A variety of tools exist which can help in the process of becoming a reflective practitioner.

Notes

1. Illeris, K. *Three Dimensions of Learning* (Malabar, FL: Krieger Publishing, 2004).
2. Ormrod, J.E. *Human Learning* (Englewood Cliffs, NJ: Prentice Hall, 1995).
3. Bloom, B.S. (ed.) *Taxonomy of Educational Objectives: The Classification of Educational Goals: Handbook I: Cognitive Domain* (New York: Longman, 1956).
4. Moon, J. *A Handbook of Reflective and Experiential Learning: Theory and Practice* (London: RoutledgeFalmer, 2009).
5. Beard, C. and Wilson, J.P. *Experiential Learning: Best Practice Handbook for Educators and Trainers*, 2nd edn (London: Kogan Page, 2006).
6. Sims, R.R. and Sims, S.J. *The Importance of Learning Styles* (Westport, CT: Greenwood Publishing Group, Inc., 1995).

23

7. Chivers, G. "Utilizing Reflective Practice Interviews in Professional Development", *Journal of European Industrial Training*, 27(1) (2003): 5–15.

8. Schön, D.A. *Educating the Reflective Practitioner: Toward a New Design for Teaching and Learning in the Professions* (San Francisco: Jossey-Bass Publishers, 1987).

9. MacFarlane, B. "Developing Reflective Students: Evaluating the Benefits of Learning within a Business Ethics Programme", *Teaching Business Ethics*, 5(4) (2001): 375–387.

10. Jensen, P.E. "A Contextual Theory of Learning and the Learning Organization", *Knowledge and Process Management*, 12(1) (2005): 53–64.

11. Hey, A., Peltier, J.W. and Drago, W.A. "Reflective Learning and On-Line Management Education: A Comparison of Traditional and On-Line MBA Students", *Strategic Change*, 13(4) (2004): 169–180.

12. Doyle, W. and Young, J.D. "Management Development: Making the Most Out of Experience and Reflection", *The Canadian Manager*, 25(3) (2000).

13. Smith, B. "Building Managers from the Inside Out: Developing Managers through Competency-Based Action Learning", *Journal of Management Development*, 12(1) (1993): 43–48.

14. Moon, *Handbook*, op. cit.

15. Johns, C. *Becoming a Reflective Practitioner*, 2nd edn (Oxford: Blackwell Publishing, 2004).

16. Ibid.

17. Hey *et al.* "Reflective Learning", op. cit.

18. Boyd and Fales (1983), cited in ibid.

19. Jensen, "Contextual Theory", op. cit.

20. Argyris, C. and Schön, D. (1996) cited in ibid.

21. Smith, M.K. "Chris Argyris: Theories of Action, Double-Loop Learning and Organizational Learning", in *The Encyclopaedia of Informal Education* (2001). Available at: http://infed.org/mobi/chris-argyris-theories-of-action-double-loop-learning-and-organizational-learning/ (accessed 8 December 2013).

22. Bostrom, R.P., Olfman, L. and Sein, M.K. "The Importance of Learning Style in End-User Training", *MIS Quarterly*, 14(1) (1990): 101–110.

23. Scott, J.L. "Awareness of Actual Learning Processes", *Journal of the Operational Research Society*, 53(1) (2002): 2–10.

24. Honey, P. "Building on Learning Styles", *Training Officer* (April 1983), Action IMC University Press – Reference LTL 141/147.

25. Kolb, D. *Experiential Learning: Experience as the Source of Learning and Development* (Englewood Cliffs, NJ: Prentice Hall, 1984).

26. Gray, D. "Facilitating Management Learning: Developing Critical Reflection through Reflective Tools", *Management Learning*, 28(5) (2007): 495–517.

27. Raelin, J.A. "Public Reflection as the Basis of Learning", *Management Learning*, 32(1) (2001): 11–30.

28. Kelly, G.A. *The Psychology of Personal Constructs* (New York: W.W Norton, 1955).
29. Moon, J. *Learning Journals: A Handbook for Academic Students and Professional Development* (London: RoutledgeFalmer, 1999).
30. Boud, D., Keogh, R. and Walker, D. *Reflection: Turning Experience into Learning* (London: RoutledgeFalmer, [1985] 2005).
31. Garvin, D. *Learning in Action: A Guide to Putting the Learning Organization to Work* (Boston: Harvard Business School Press, 2000).
32. Williams, T. "Learning from Projects", *Journal of the Operational Research Society*, 54 (2003): 443–451.
33. Chivers, "Utilizing Reflective Practice Interviews", op. cit.

CHAPTER 4

Different types of learning for the HSE practitioner

Initial professional education

Entry into the health, safety and/or environmental management profession follows a model used by a number of professions. In the UK, National Occupational Standards (NOS) exist for a wide range of occupations. These NOS are statements of performance that describe what competent people in a particular occupation are expected to be able to do. They cover all the main aspects of an occupation, including current best practice and the knowledge and understanding that underpin competent performance.

Sometimes the core competence standard is developed by a professional body which then uses that standard to evaluate the curriculum from universities or qualification-awarding organisations and approves them as meeting their entrance requirements for a grade of membership. Sometimes the professional body itself awards qualifications as well as membership.

This is the model followed by a number of professions, including engineering and human resource management. Unlike certain professions, the title of Occupational Safety and Health and/or Environmental Practitioner is not one reserved for members of the professional body, though use of postnominals (literally, letters after your name) is reserved, and in some countries it is a criminal offence of fraud to use postnominals that you are not entitled to.

In the traditional education of occupational health, safety and environmental (HSE) practitioners, a syllabus is developed based on the NOS or a vocational standard agreed by a group of eminent professionals who agree on what a competent practitioner needs to know and be able to do.[1] In some cases, safety, health and environment are dealt with as separate disciplines.

Development of professional standards increasingly happens across national boundaries. In Europe, the European Network of Safety and Health Professional Organisations (ENSHPO) has a Certification Standard for European Occupational Safety & Health Managers. Safety professionals in Europe who are already a member of a professional body at a specified level can apply to use the formal title of "EurOSHM". Those who may not be a member of a professional body, but who meet the education and experiential criteria can also apply.

The International Network of Safety and Health Professional Organisations are undertaking a similar exercise to identify the core curriculum across national boundaries. However, this work is focusing on what the national professional organisations agree that safety practitioners need to know and are able to do now. It is more difficult to consider what they will need to know and be able to do in the future.

The broad agreement on the cognitive learning needs of practitioners includes content on specific hazards – the technical knowledge, and techniques of management such as assessing risk or managing change. For example, the European Criteria for Approved Curriculum[2] is reproduced in Table 4.1.

Options for initial professional development

Acquiring the body of technical knowledge to be an effective safety practitioner can be achieved through a number of routes. The choice of route needs to be made by the individual. There is no one "best" route: each has its own advantages and disadvantages.

Many professions have entry at age 18, commencing with the course of study followed at university, for example, medicine and engineering. Others have a well-recognised postgraduate entry scheme, commencing around

Table 4.1 List of the indicative subjects that could be included in the required training courses, ENSHPO Certification Standard for European Occupational Safety & Health Managers

	Subject area	Indicative subjects
1.	European and national occupational safety and health regulation	• Relevant European safety and health legislation and its translation into national practice • Regulatory mechanisms relevant to occupational safety and health in the public and private (civil law) spheres influencing and responding to regulation • Occupational safety and health in the context of public policy
2.	Safety and health management	• Setting and improving policy for occupational safety and health • Organising for safety and health; safety and health management systems • Safety and health auditing • Organisation of the protection and prevention services • Promotion of a positive safety and health culture • Management of contract works • Monitoring, reviewing and auditing of health and safety performance • Basics of environmental management
3.	OSH risk assessment and management	• Risk assessment methodologies and implementation • Risk management (identification and successful implementation of specific risk control measures) • Developing safety methods of work, safety instructions, etc. • Best practice
4.	Occupational safety and health technical knowledge	• Accidents and occupational diseases investigation, recording and reporting • Occupational safety science (for example, machinery and work equipment safety; electrical safety; construction safety; fire safety; accident prevention techniques, working at heights) • Occupational health and hygiene science (for example, chemical, physical and biological hazards and exposure limits and prevention measures)
5.	Safety training, information and communication	• Safety and health communication techniques • Training assessment, execution and evaluation
6.	Human and ergonomic factors	• Posture, manual handling and musculoskeletal disorders; anthropometry and work physiology • Workplace design and layout, including computer workplaces • Human behaviour and safety
7.	Advisory and change management skills	• The OSH manager as change agent • Organisational learning • Technical and organisational change management
8.	Project work	• The course should provide the opportunity for the course members to apply the lessons learned in theory to the practical situations in their own or other workplaces and to report on that process

Source: ENSHPO (2013).

age 21 where the entrant's first degree may or may not be related to the profession. An example of this is law.

While there are undergraduate degrees in safety, health and environmental management, most entrants to the profession tend to move into this career later in their working life. Sometimes they will have made considerable progress in another career, and take on safety as an ancillary to their main role before moving into it as their full-time role. Research conducted by NEBOSH on the demographic characteristics of those undertaking their professional level Diploma qualifications reveals an interesting picture: 39 per cent of the students taking the International Diploma in the year to March 2013 had a Master's level qualification or higher and 83 per cent were in the age range 30–49.

National Vocational Qualifications (NVQs)

National Vocational Qualifications or NVQs are based on the concept of workplace learning. In most cases, the learner develops a portfolio of evidence of their competence to carry out the prescribed activities within a National Occupational Standard.

Award of the qualification by the Awarding Body is made via an assessor working for the approved training organisation (which in some cases can be the person's employer). The assessor's view is in turn verified by an internal verifier working for the training organisation, with maintenance of standards and quality control undertaken by an external verifier working for the awarding organisation.

The assessor's role is important throughout the process. They work with the candidate to do the following:

- establish what the candidates can already do;
- agree on the standard and level they are aiming for;
- identify what they need to learn;
- agree the activities the learner will undertake in order for this learning to take place.

Candidates might take a formal course if that is the best way to learn what they need, or they might negotiate a work placement or project with their employer to gain evidence of a competence. Identifying the learning requirements with the assessor is in many respects a reflective conversation, requiring the trainee to reflect on their competence and plan an activity. Carrying out that learning activity, recording it in the portfolio and the subsequent conversation with the assessor to reflect on whether the competence requirements have been met can be considered to be stages in Kolb's learning cycle (see Chapter 2).

The assessor will "sign off" the achievement of a unit within an NVQ when they are satisfied that the candidate has met the required standard of competency. In each case they are looking for evidence of that competency. The evidence may be examples of written work, such as risk assessments; it can be photographic, audio or video evidence of the person actually demonstrating a skill. It could also be the assessor's direct observation of the candidate, or questioning of the candidate to assess underpinning knowledge and understanding. The assessor will also question the candidate to authenticate the validity of the evidence, for example, that it is their own work within the portfolio.

There are five levels of NVQ, ranging from Level 1, which focuses on basic work activities, up to Level 5 for senior management. The NVQ in health and safety is available at levels 3, 4 and 5.

Some advantages of NVQs are that the learner is able to relate their learning immediately to their own workplace. They are often able to make a contribution to workplace safety while completing the qualification, by their activities of policy development, risk assessment and evaluating control measures. This can be seen to add value to the employer.

Where the NVQ is rigorously assessed, the learner will have successfully demonstrated their ability across the whole range of competencies required to successfully fulfil that job role. The NVQ format minimises the time that the learner is away from the workplace, and can therefore often suit the needs of more mature entrants to the profession who may be combining their studies with full-time employment and family responsibilities.

Some individuals really do not demonstrate all they are capable of in an examination, which requires focused consideration of what is required to answer a specific question and the construction of an answer in a limited time period. It is arguable whether examinations simultaneously test a different skill set, including response to pressure, recall of information and the ability to articulate the application of recalled information to the subject of the question. NVQs may suit those who dislike examinations.

NVQs suit certain learning styles. Activists in the Honey and Mumford model are likely to prefer learning by doing and enjoy the practical aspects of an NVQ. It can be argued that the process of evaluating yourself, with the aid of your assessor, against a standard of competency, identifying learning needs, carrying out activities to meet those needs and then assembling evidence of competency in that area, actually enhances the learner's ability for reflection and future reflective learning. Conversations with the assessor can be regarded as akin to working with a coach or mentor as a reflective learning technique. It also opens the individual up to the possibility that not all learning is based on attending a course. Activities in the workplace, working with other parts of the organisation and with different people can be valuable learning opportunities when linked to reflection.

There are, however, also some disadvantages to NVQs. In the UK they were introduced in the mid-1980s, so they are a relatively recent addition to the qualifications landscape. For this reason they may not be well understood by all organisations, especially since those presently in senior positions are likely to have come up through the formally examined qualifications route. NVQs in Safety and Health were available from the mid-1990s onwards, so it is still relatively early to evaluate the impact of this type of qualification on the subsequent holder's career path compared to holders of degrees or diplomas in the same subject.

NVQs sit within a framework of qualifications with each qualification designated at a level. Level 4 and above correspond to professional level qualifications. Over the five years from 2007 only 1.6 per cent of all vocational qualifications awarded in England, Wales and Northern Ireland (including NVQ-style qualifications and vocationally related qualifications which may be assessed by examination) were at level 4 or above.[3] Some 85 per cent of vocational qualifications are awarded at level 2 or below. Where employers are familiar with NVQs, it is likely to be for practical, skill-based occupations rather than for professional staff.

The development of the portfolio and the organisation of evidence to satisfy all the criteria within the NVQ standard can be administratively demanding, and this process will be more challenging for some learners. There is also a danger that the portfolio will simply become a mountain of paperwork.

One further issue which has negatively impacted on employers' perceptions of NVQs is that the assessor is likely to be involved in the learning process of the learner, and hence may not be impartial in making their awarding decision, notwithstanding the quality control processes from internal and external verifiers. The process of external verification involves sampling assessment and verification practice, providing feedback and an External Verifiers (EV) report to the centre and the appropriate Awarding Body.

The verification chain, specifications and evidence of achievement were viewed by Eraut and Steadman,[4] in their evaluation of NVQs at management level, to be paper-dominated, and ultimately not necessarily guaranteeing standards and fairness.

Typically, candidates will work towards an NVQ that reflects their role in a paid or voluntary position. The implication of this is that it is very difficult for someone not already in a safety-related role to achieve an NVQ in that field.

Nevertheless, despite the detractors of the NVQ system, the authors would stoutly defend NVQs in health and safety. Coming from a relatively privileged background of being able to go to university at age 18, and having opportunities not available to others of our generation, ethnicity and gender, we welcome the opportunity for people to develop in later life and make a contribution to the profession.

Vocational related qualifications

A vocationally related qualification, like an NVQ, is developed from the National Occupational Standard, where one exists, but may go beyond it in specifying the content of the course of learning or syllabus.

It is important that syllabuses are reviewed regularly because new working practices and processes develop and risks change as a consequence. An example is in the development of machinery and production technology. In the early and mid-twentieth century, many factories and workshops had a range of machinery powered by a prime-mover and a series of transmission machinery, such as rotating shafts and belt drives. This system led to hazards of entanglement at multiple locations around the factory or workshop, safeguarded – where indeed such safeguards were present – by fixed guards. Although machinery could still be found powered in this way in older factories both in the UK and in developing economies in the late twentieth and early twenty-first centuries, the student of machinery safety needs to also understand the hazards arising from the use of robots in production.

An example of an increased risk which has arisen in the past 30 years is a consequence of the high prevalence of Human Immunodeficiency Virus (HIV) in Sub-Saharan Africa. In an occupational context, in countries with a low prevalence, the main concern is avoiding contact with waste body fluids which can cause infection, for example, to healthcare workers. However, if a high proportion of the workforce is HIV positive, that completely alters the risk profile of the organisation. For example, a simple cut can have severe consequences in individuals with a compromised immune system. There are two significant implications here for the learning needs of those within the health, safety, or environmental profession. The first is that it needs to continue throughout a person's career because new technology introduces both new hazards and new ways of controlling those hazards. It is almost impossible to imagine the changes in technology that will occur over the course of a 30- or 40-year career. However, learning to reflect and learn from what you do every day can help you to change and adapt to the challenges of the future.

Second, in moving between industries and even countries in the course of one's career, cultural norms and working practices can have a bigger impact on risks than the technology in itself. While information on the hazards of an industry and the control methods is relatively easy to find, the challenges of implementing those technically rational solutions in an organisational context can present the most significant problem to the practitioner, but also perhaps the greatest learning opportunity. If you fail to learn to influence those around you, you fail to become an effective safety, health or environmental practitioner.

Vocational qualifications will usually involve following a course of training, and then demonstrating learning by means of an assessment. The assessment is frequently a written examination and a workplace-based assessment demonstrating the ability to apply what has been learned to a real situation.

An advantage of this traditional mode of professional training is that delivery of the course can be very flexible. Some students attend classes, sometimes delivered in evenings or weekends to fit around the working week. Increasing numbers of students access training through distance learning, sometimes over the internet. This can make the qualifications accessible wherever in the world the student happens to be based. It allows students to achieve the qualifications without the support of an existing employer.

Examinations are seen by many people as being an academically rigorous form of assessment. Examinations can be set at an appropriate level so that they do not merely require the regurgitation of facts and information, but require the candidate to demonstrate their ability to apply knowledge, to evaluate a given scenario and to propose a solution, following the levels shown in Bloom's *Taxonomy* discussed in Chapter 3. There is also clear separation of the training delivery and the assessment, in that the tutor, who has a relationship with the learners, does not participate in the assessment of those learners. Setting and marking of the examination are undertaken by an impartial independent awarding organisation, regulated by the government or a government agency.

Employers tend to value vocational qualifications and find them more understandable, since this is likely to be the route that many in senior positions have followed themselves.

However, this form of qualification and assessment is not without its disadvantages. There may be some scope in the practical assessment to apply learning to the specific issues in the employee's workplace, but in general there is not the same opportunity to study the specific issues in that workplace as the vehicle for the qualification as there is with an NVQ. Instead the learner will study a range of hazards which will give them the tools to approach evaluation and control of the specific hazards within any workplace. It can be argued that this means that the learner is more flexible and can apply their skills in many occupational contexts, but it does mean that there is a leap which has to be made by the learner to put their education into practice.

Schön describes a "practicum" within the training of many professionals.[5] That is an environment in which they can develop their skills by "doing", within a learning environment. For example, in medicine, junior doctors learn to make a diagnosis with real patients, but initially under the close supervision of an experienced colleague who can coach and mentor through the process. In legal practice, trainee barristers undertake a pupillage with one or more experienced "Masters".

The second Master I had . . . she was taking me around places and talked to me about cases in a lot more detail. So this is before I started practice. In the first six months I was shadowing . . . So she would explain things and sit me down and get me to reflect on the facts of the cases and say to me: "Well, what would you do in these circumstances?" And sometimes I was right and sometimes I wasn't and when I wasn't, she told me. You know, and that helped a lot to shape my practice when I actually started appearing in court in my own right.

HSE Consultant and Barrister

Many HSE professionals are the sole practitioner within their organisation, and do not have the opportunity to be guided in this "leap" by more experienced colleagues or a coach. When they do work alongside others to help negotiate this transition to real-life practice, most find it hugely helpful.

They gave me a project on environmental impact assessment but I was working with a senior who has a lot of experience . . . So for this project I reported to that senior and . . . conducted the environmental impact assessment under his [supervision]. I learned a lot about environmental impact assessment, the things which I didn't grasp in my initial training.
Senior Environment Compliance Officer, Oil and Gas Sector

Traditional examinations also need to be appropriately invigilated or proctored in order to guarantee their rigour. Malpractice in examinations is quite rightly condemned in most parts of the world. However, where holding a particular qualification can greatly enhance earning ability, the temptation to cheat to pass can be strong. More blameworthy still are course providers who assist in this process out of misguided loyalty to their student or for financial reward.

"Rote learning" can be a serious issue in professional training. Where knowledge is tested by examination, there can be a temptation to learn key knowledge or model examination answers, word for word, without real understanding of the meaning of what is being "learnt". Both malpractice and rote learning are usually apparent in examinations scripts and Accredited Awarding Organisations will have mechanisms in place to detect such malpractice and deal appropriately with offenders.

University degrees in Health, Safety or Environmental Management

It is possible to take a degree in health and safety both at an undergraduate or bachelor's degree level, and at postgraduate or higher level degree such as a Master's. Sometimes, those achieving a vocationally related qualification which is at a comparable level to a bachelor's degree, having already achieved membership of a professional body at an appropriate level, will go on to take a Master's degree which may specialise in an area of particular interest to them, may be technically related to their job role, or may give them the opportunity to pursue research into a safety, health or environmental management-related topic.

Achieving professional membership

Admission to a professional body additionally usually requires evidence of certain experiences and a commitment to continued professional development and adherence to a code of ethics or professional practice. Additionally, most professional bodies will provide opportunities to continue to learn and develop their members' professional practice by access to journals, conferences, training events and networking meetings, often including a formal presentation by another member.

Continuing professional development (CPD)

Some professional bodies manage their continued professional development (CPD) on the basis of the accrual of a certain number of "points" achieved over a cyclical time period. Points can be allocated on the basis of the time spent on the development activity. However, this can lead to development activities being undertaken simply to achieve points, and which may be unrelated to the actual learning needs of the practitioner. A better approach is to consider continued professional development in light of specific learning needs, and the processes of reflective practice can be helpful in identifying those needs.

We have already briefly discussed how a change of industry, location or technology can give rise to a need to update skills and knowledge. However, some events in a working life can give rise to a need to rethink the approach taken to the job role, and – if the opportunity is taken – to learn new ways of achieving results. This will be discussed further in Chapter 5.

It is a sad fact that many health, safety and environmental professionals equate professional development with attending training courses. Training courses certainly have their place, but there are numerous other methods of acquiring knowledge, skills and behaviours which can enhance performance in a professional role. The key issue is that the individual must not be the "passive" recipient of professional development. He or she must take responsibility for identifying their own learning needs and determining the best way to meet them. Even if that method is to attend a formal course, there is work to be done back on the job, to reflect on how the learning can be put to use in the individual's own context.

Professional development opportunities

Formal training courses

There are numerous formal training courses held on health, safety, and environmental topics. Some courses can prepare participants for further qualifications. This may be extremely useful where the scope of a person's role increases to take in new disciplines, for example, a safety practitioner who then becomes responsible for environmental issues, or vice versa. Other formal training courses deal with specific hazards and their control, or a management technique such as root cause analysis, environmental impact assessment, using hazard and operability studies or introducing a behavioural safety programme. More general management training courses can also be invaluable to HSE professionals, such as interviewing techniques, project management or presentation skills. Courses may be run by commercial training organisations, universities or colleges or by a professional membership body. Courses can be available through on-line learning programmes as well as by attendance at a traditional "face-to-face" training course.

Receiving coaching or mentoring

Coaching is usually a one-to-one relationship with a more experienced colleague. Coaching can assist the practitioner in determining different ways of approaching problems. The role of the coach is quite crucial, and many skilled coaches can flex the role they adopt to suit the needs of their "student". For example, they may resist providing the practitioner with the "correct answer" and instead seek to help them explore a range of options which may achieve the desired outcome.[6]

Mentoring can be a more informal process arising from relationships with more senior people perhaps in other areas of the organisation or in a professional body. This can be of particular value for those HSE practitioners who are the sole practitioner in their organisation.

There was a couple of guys when I was working at [the training organisation] that were older practitioners with other companies that took me under their wing . . . and they kind of pulled the strings a little bit, you know, laying down requirements, and I got on with both of them, so I did a couple of investigations with each of them, and I did learn quite a bit from them. So that was nothing formal – I just took the opportunity.

Health and Safety Consultant and Barrister

Reading relevant books, journals and reports

It is important to keep up to date with current knowledge and best practice in the HSE field. There is no shortage of journals and on-line resources to enhance your knowledge and understanding, and these are frequently more topical than textbooks which take a while to publish.

Research

Research is a useful tool both in improving HSE practice, both within an organisation or an industry sector. It is also an invaluable tool for personal professional development.

For the individual practitioner, it could include in-depth analysis of a pattern of incidents within the organisation to identify underlying causes. It could include speaking to others in the same industry sector to determine the solutions they have implemented to mutual problems. It could be undertaking a survey of employees' views or beliefs on a specific issue, for example, to determine the effectiveness of communications on HSE topics. Research can provide answers to problems and enable practitioners to implement changes to their own practice.

Working collaboratively with other professionals in the same sector on research can also be fruitful. Pooling of data across a number of organisations can provide clearer insights into relatively rare hazards or incidents. Sometimes joint research can be facilitated by an industry body, with the benefit to those participating of sharing the outcomes. It could involve commissioning research, for example, from a university or from the Health and Safety Laboratory in the United Kingdom to identify better solutions to common industry problems.

In a more formal context, it is possible to achieve a Master's degree or doctorate based on research conducted in the workplace.

Visiting other organisations or other locations of your organisations

Networking with other professionals in the same field can be an extremely useful professional development technique. This can extend to visiting each other's workplaces to see first-hand how particular issues can be dealt with. It can be of enormous benefit to those undertaking a formal qualification, where the topics in the syllabus are not part of the working experience of an individual learner.

Attending meetings

Most professional bodies hold meetings of their "branches" or "chapters" at regular intervals. Often these will include an external speaker on a topical issue, as well as opportunities for networking with other members. Some sectors will have formal or informal meetings of HSE professionals within that sector. Simply networking at a meeting with other professionals should not be overlooked as a means of professional development. Side conversations can often lead to insights into improvements in an individual's practice. Building a network of contacts can mean that when you have a complex query, you know a person that can advise. If a network does not exist – then set one up!

Internal meetings can also be learning opportunities, particularly where the purpose of the meeting is to work out solutions to problems or deliver a project.

Writing for journals or books

Writing for a HSE journal can be a continuing professional development activity. It can provide an opportunity to disseminate lessons learned or case studies from your own practice. It can enhance the author's own learning by requiring additional research and by enabling reflection on the issue. Even if you don't succeed in getting your article published the first time, the writing will be a useful reflective learning experience and the feedback on how to improve next time is worth some CPD points.

Work shadowing

Spending time with other professionals can be a learning experience. It may be useful to spend time with a more experienced colleague dealing with a situation or hazard with which the learner is unfamiliar. It may be useful simply in observing how a colleague deals with difficult situations. This gives rise to a learning technique called "modelling". This can be more simply described as imitating what works for someone else. Children "model" behaviours from what they observe from the adults around them.

My initial job was going round the schools and technical colleges testing fume cupboards and dust extraction systems. They sent me on a course which was basically three days and I felt I didn't really know enough about it after this time. I made enquiries of various companies that did testing and heating and ventilating companies . . . and I went out with their guys.

. . . because the three day course they sent you on, you go into this LEV system and there'd be holes here, you'd put your pitot tube in there and you'd take the reading.

No, that's not what it's like in real life. You go in this place that's smothered in dust, the stuff is up out of reach. You've got to climb on ladders; you have to drill holes yourself; you have to decide where to put the holes. So I went out with the guys that were doing it and got some input from them as to how to do it.

Health and Safety Manager, Local Authority

Peer-to-peer evaluation

This technique involves evaluation of the skills of an individual by their peer with feedback provided. This can be useful as a learning technique both for the person being evaluated, and for the person giving the evaluation. There are some important considerations to make this effective. First, it should be an equal relationship, and both parties must enter it with an attitude of learning and acceptance of the feedback in a spirit of constructiveness. The person making the evaluation should be a "critical friend" or trusted colleague.

Working on projects

A project is a defined time-bound piece of work designed to deliver a business benefit. It often involves a group of people – the project team – brought together from a range of disciplines to bring their expertise to the work of the team. Successful projects are characterised by clear definition of the outputs to be delivered from the outset, and careful planning and co-ordination of the tasks, activities and resources to achieve the end result.

Project management, or programme management, which is the co-ordination of a number of projects to deliver the required change, is a professional discipline in its own right. It is also a competence that some safety practitioners need to develop where they are required to bring about complex change. An example of where a project management approach would be appropriate is the introduction of a new software system to report accidents through a business. This would not be an activity that is simply undertaken by the "IT Department" though they would almost certainly have a part to play. The activities that need to be co-ordinated and the products to be delivered are likely to include identification of the business need; the selection of a system from a reputable supplier that will continue to be supported and developed as operating systems are updated; the specification and configuration of the system; the specification and purchase of additional hardware; the installation on computers; the production of a training package for users; the implementation and recording of training; and then systems for on-going support to users; and monitoring of compliance with the system's use. Involvement of users will be critical to the success of many stages of the project and its ultimate success in delivering the business benefit.

Project management, even if the safety professional is delivering a less complex project than that described above, is an essential skill in bringing about change and improvements. It could be argued that this is at the heart of the safety practitioner's job: his or her reason for being.

Formal project management systems and training in those systems are widely available. Bridging the gap between the theory of managing a project and the actuality can be more of a challenge. Volunteering to work on a project, and maximising the learning opportunities it presents can enhance not just competence on managing projects, but also a range of other valuable skills.

Working in a multi-disciplinary team and learning from the key competencies within their roles – their "day jobs" can be very useful as described above in the section on work shadowing. Observing first-hand the approach, tools and techniques of a project manager can be learning that is transferred directly into practice the next time the safety practitioner has to implement a change. The complex example described above can give rise to opportunities to learn more and refine skills in communications, listening and understanding user requirements, making purchasing decisions, selling benefits and techniques of persuasion, and managing the emotional journey that many people go through – which is similar to a grief cycle – when required to deal with change in the workplace.

Participating with the project group can enhance skills in team-working, time management and prioritising. Finally, a good project will conclude with not just a celebration of success, but with a "lessons learned" meeting. This is an example of group reflection, and the opportunities can be taken to consider not just what could be improved upon, and the suggestions for what can be done differently, but also reflecting on what went well, why it went well, what aspects of it worked best, and the learning points that can be taken forward. It is a golden opportunity to move out of your comfort zone and see the world through other people's eyes – and this is a key skill in becoming a reflective practitioner.

Undertaking community projects

In the UK-based Institution of Occupational Safety and Health (IOSH), one of the criteria for moving into the membership grade of "Fellow" is to have made an impact on health and safety beyond your own workplace. Many HSE practitioners do volunteer their advice in other contexts, perhaps based on a sport or hobby that they undertake, or advising on safety in events within their own community. There is a lot to be gained from these activities, which expose the practitioner to a completely different dynamic than exists in the workplace. For example, it may be far easier to persuade work colleagues to adopt particular working practices, than fellow volunteers – a challenge which can hone your interpersonal skills.

Volunteering for committees

If you are a member of a professional body, there is usually a branch or chapter structure, largely run by volunteers from within the membership. Within the body as a whole there may be a number of committees on which volunteers can serve. Volunteering for the committees undertaking various activities can be rewarding and an amazing learning opportunity. For example, local branches or chapters often have a programme of events or external speakers. Organising these events is a skill which can be transferred back into the workplace. The choice of speaker and topic can also address the learning needs of the branch membership. In fact, not only can the topic provide valuable information, but reflecting on the presentation style can also be a learning opportunity in itself. In other words, if the speaker was entertaining, engaging and communicated well, it can be useful to reflect on what it was they did that made it so engaging, and resolve to try some of the techniques in your next presentation. Likewise, if the presentation was dull, what did the speaker do wrong? Are you guilty of some of the same faults?

Learning from other professions

Many of the skills which need to be exercised by the competent HSE practitioner are the expert province of other professionals. It therefore follows that skills and expertise in these areas can be learned from other professions. A few examples are set out below. If these job roles exist within your own organisation, then the opportunity can be created to spend time learning from your colleagues. If they are outsourced to other organisations, it can be more difficult because you may be racking up "billable hours".

Most people are very willing to help their colleagues even in unexpected areas. You can learn from colleagues by observing them at work, or observing the outputs of their work and discussing with them why they took the approaches they did. You can also ask them to observe you at work, for example, delivering a presentation or conducting an interview, and provide you with honest feedback on areas to work on and techniques that you might try. You can ask for feedback on finished pieces of work such as reports of investigations. All these approaches will give you fuel for your own reflection and ideas that you can try out to be reflected on further as you build your skills and competence as a HSE professional.

Beware of "cut and paste" solutions. You are seeking to develop your own skills, not to plagiarise others' work. In fact, there can be a danger that using others' work as a template for your own discourages reflection and hence wastes the opportunity to extend your own repertoire of skills and competence.

Interpreting legal requirements

Larger organisations may have their own in-house legal team. Considering the approaches they take in developing policy when interpreting general legal requirements as they then apply to the whole organisation can be illuminating and can inform your own work in developing policy dealing with the organisation's response to HSE laws.

Auditing

The Health and Safety Executive in the UK define auditing as: "The structured process of collecting independent information on the efficiency, effectiveness and reliability of the total health and safety management system and drawing up plans for corrective action."[7]

The key point is that auditing involves systematic evaluation of a system, usually against a standard. Auditing on other topics within an organisation may be carried out by accountants, an internal audit function and quality professionals. The correct methodology of auditing can be learned on formal training courses, but deeper understanding and honing of skills can be achieved by shadowing or practising under the supervision of a quality auditor.

Developing policies and procedures

There are a host of skills involved in developing policies and procedures. Quality professionals may have sound advice on these issues, but other colleagues may also be involved in writing procedures which are unambiguous, meet legal requirements and are readily understandable by those that need to use them, for example, colleagues in Human Resources.

Interviewing and investigation

Interviewing skills can be improved by formal training. It can often feel intrusive to probe during interviews to pin down specific facts, but in conducting an investigative interview, it is important to be very clear on what a witness is telling you as opposed to jumping to a conclusion about what they are trying to say. Colleagues who have worked in the police or fire services or in health, safety or environmental enforcement services who have been involved directly in interviewing witnesses following an incident may have very developed skills in this, and observing them can be illuminating. You could also ask them to sit in on an interview that you are conducting and give you feedback on your technique. Colleagues in HR may also be involved in conducting investigations on disciplinary matters or in recruitment and again may have skills that you can learn from them.

Report writing

Writing a report following an audit, an inspection, or an investigation, or preparing a report to persuade others to adopt a course of action is a demanding task. Professionals who need to do this well include journalists and quality professionals, among others. Discussing the principles with them and asking for feedback on your own reports can be helpful in honing these skills.

Delivering training

Many practitioners avoid giving training at all costs. Speaking in front of others is a fear affecting many people in all walks of life. However, at some stage, whether it is a formal training course, or a presentation about

HSE performance within the business to a meeting of senior managers, presentations skills will need to be mastered. Training professionals should obviously have these skills and may pass on some useful tips and skills. If you ever get the opportunity to discuss presentation with an actor, they can also give you tips and techniques for commanding and retaining the attention of your audience.

Communications

Larger organisations may have a communications department who can assist in drafting documents so they can be readily understood. There are greater challenges in workplaces where a variety of languages are spoken and many people are not operating in their first or native language. Drafting documents for consumption by people who are operating out of their first language often requires simplification of the language and the avoidance of idiom. Translation, particularly of technical documents, can be very difficult, where the translator may not appreciate the meanings of technical expressions like "permit to work". Often this expression is used as it is across a number of languages, and a literal translation would be misleading and inappropriate.

There may also be problems with literacy, and a communications department can give advice on how to overcome this to ensure you are giving consistent messages.

Making a business case

Cash is a finite resource. Whatever you choose to do with cash, you are choosing not to do something else. In making a decision about allocating budget or capital expenditure to a project, managers are seeking to maximise returns or the potential benefit to the organisation. Even within the not-for-profit sector, managers are accountable for the responsible stewardship of corporate assets. Setting out the costs and potential benefits of a proposed course of action is a key part of achieving agreement from those responsible for allocating budgets.

Many professions need to make a business case, and hence have expertise that the HSE practitioner can learn from. Examples include project managers and managers in finance, product development, IT and purchasing.

Conclusion

There is a plethora of different types of learning available to HSE practitioners, from your initial training through your entire career. There are choices to be made regarding your initial professional development, which depend partly on your personal preferences and also what is realistically available to you. However you go through your initial professional training, it is important to give yourself the time to reflect. Ideally, you should do the following:

▶ Consider how your learning links and connects with or contradicts what you already know.
▶ Formulate ideas about how you can put your learning into practice in your work.
▶ Identify the parts of your learning that you are struggling with and consider alternative ways of exploring those topics.

Learning has to continue throughout a career as organisational and hazard contexts change. Learning to be reflective can enhance the opportunity to learn "on the job" and build your confidence and competence as a practitioner.

There are numerous opportunities for development quite apart from those formally provided through training courses. Reflect on your needs, your areas of difficulty and critical events that occur and proactively identify opportunities to enhance your skills.

Consider your colleagues both within and outside of the profession as a rich untapped resource for your own development. Often the skills other professionals develop are of direct benefit to your working practice and they will be happy to assist you in learning from their experience.

Finally, don't be selfish. Always be willing to share your knowledge and skills with new entrants to the profession.

In brief

▶ There are many ways of completing initial professional development in HSE. No one method is inherently better or worse than any other and the choice depends on resources, opportunity and the preferences of the new practitioner.
▶ Continuing professional development (CPD) is vital in maintaining competence as well as being a mandatory requirement of most professional membership bodies.

▶ CPD is not just about attending courses. There are numerous opportunities to develop your competence and professional practice even with limited resources.

▶ Many of the skills which need to be exercised by the competent HSE practitioner are the expert province of other professionals. Build your network of mentors both inside the HSE profession and outside to maximise your learning opportunities.

Notes

1. National Examination Board in Occupational Safety and Health, *NEBOSH Guide to the Diploma in Occupational Health and Safety* (Leicester: NEBOSH, 2008).

2. ENSHPO Certification Standard for European Safety and Health Managers (EurOSHM), May 2013. Available at: http://www.euroshm.org/full.php (accessed 15 December 2013).

3. Office of Qualifications and Examinations Regulation (OfQual) "Regulated Qualifications Activity Dataset 2007/8 to Present". Available at: http://ofqual.gov.uk/standards/statistics/raw-data (accessed 15 December 2013).

4. Eraut, M. and Steadman, S. "Evaluation of Level 5 Management S/NVQs Final Report 1998", Research Report Number 7 (Brighton: University of Sussex, 1998).

5. Schön, D.A. *Educating the Reflective Practitioner: Towards a New Design for Teaching and Learning in the Professions* (San Francisco: Jossey-Bass Publishers, 1987).

6. Ibid.

7. Health and Safety Executive *Successful Health and Safety Management*, HSG65, 2nd edn (London: HSE, 1997). Available free to download at: http://www.hseni.gov.uk/hsg65_successful_h_s_management.pdf.

CHAPTER 5

The importance of reflective learning for HSE practitioners

10,000 hours . . .

There are a number of recently published books discussing the nature of expert performance. *Bounce: The Myth of Talent and the Power of Practice* by Matthew Syed[1] is a book written by a British table tennis champion. Fêted for his talent, his exploration of the topic was triggered by the knowledge that a generation of British table tennis champions lived within a few streets of him. Clearly there was something in the environment which led to this success, building upon the raw talent. His conclusion? 10,000 hours of practice, but practice of the right type.

Guitar Zero: The Science of Becoming Musical at Any Age[2] is the journey of a psychologist and self-confessed wannabe rock star who took a two-year sabbatical to test out the theory that 10,000 hours of purposeful practice could transform him into an ace guitarist.

Similar theories are propounded by Malcolm Gladwell in his book, *Outliers: The Story of Success*[3] and are based on work by psychologists Anders Ericsson, Krampe and Tesch-Romer.[4]

Rasmussen's concepts of skill-based, rule-based and knowledge-based behaviours are well known to many safety practitioners, underpinning as they do the theories on types and causes of human error elaborated by James Reason.[5] Skill-based performance is behaviour which can be described as a routine which does not require conscious control. It is something that you are so accustomed to doing that you do not need to think about it, or pay attention to it. It is run off as a smooth sequence. Everyday life examples are steering a car, changing gear and writing your signature.

As you become more "expert" at an activity, you tend to move from knowledge-based behaviour, where it very much needs to be under your conscious control, requiring a lot of attention, through rule-based, to skills-based.

The 10,000 hours theory reinforces this. In his book, *Bounce*, Syed uses the example of a champion tennis player. The true champions "read" and react to where their opponent is going to place the ball before the racquet strikes it, based on their experience. This process happens below their level of conscious control.

The 10,000 hours practice needs to be with some purpose. Syed describes a new coach working with him on his table tennis forehand so that it was always consistent. With this greater degree of consistency, he was able to predict and control where he placed the ball (interestingly, he was re-modelling a skills-based piece of behaviour, which has actually developed a neurological pathway in the brain. It takes a great deal of practice and repetition to change this behaviour into a new skills-based routine.)

Syed describes this as purposeful practice. It means that you are focusing your practice on the areas you need to improve, not merely repeating the things you already do well. It is akin to reflecting on what you need to be able to do, trying it, reformulating your theory and testing it further. Reflecting on what works then polishes your approach until some of your behaviours become practised and natural. As you progress and build up your expertise, you are then able to draw on your repertoire of practised skills as the need arises.

Expert performance in the professional's working life is similar. Experience leads you to read and react in light of your expertise and your previous encounters with similar problems.

The HSE practitioner and reflection

What has this to do with the work of the HSE practitioner and reflective practice? Currently throughout the world, there is a shortage of very high calibre HSE practitioners. This is partly due to the fact that this is a very young profession when compared with the traditional professions such as medicine, law, or engineering. In the UK, there were a number of events which precipitated a surge in entrants to the profession. In 1975, the Health and Safety at Work Act came into effect in UK law. As a direct consequence, the first safety and hygiene department was established at the University of Aston in 1976, and an undergraduate degree in safety ran from 1977, providing an influx of new graduates. Further changes in the law in 1988 with the Control of Substances Hazardous to Health Regulations led to a new tranche of entrants, particularly from among those already working in occupations or for companies related to chemistry. In 1992, when the first wave of legislation from Europe was introduced, increased demand for the "competent person" required to advise companies on meeting their obligations under the new law, led to a new intake of entrants.

Many HSE practitioners have not actually studied safety or environmental engineering or sciences as a base degree or diploma. Many have risen through other technical disciplines, perhaps with a safety element, or have exercised that discipline in an industry sector with the need for considerable control of high hazards.

> *I graduated from university with a bachelor's degree in chemical engineering so my background is chemical engineering . . . When I was studying, we had lots of projects related to the environment. My graduation project was on cleaning up oil spills from the sea which I chose because I was so interested in the environment . . .*
> **Senior Environment Compliance Officer, Oil and Gas Sector**

Others come into health and safety through very unlikely routes.

> *I gradually worked my way up until I was herd manager on a 300-cow herd.*
> **Health and Safety Manager, Local Government**

Safety, health and environmental management are most often second or third careers for practitioners. This is now changing, but the average experience of safety/environmental practitioners is substantially less than those of a similar age who are working in production, operations, marketing, law, education or medicine. To retain credibility, the HSE practitioner needs to learn quickly. Further, the skills that the practitioner needs to develop are complex and require significant interpersonal and change management skills as well as technical problem solving ability.

Unfortunately HSE does not have a formal training route where practitioners have a structured programme to learn both the underpinning knowledge and the practical skills, as is the case with medicine or law.

There was a lot of trial and error and what I had to rely on really was my experience from an engineering background and the leadership skills I had up to when I was about 30 to be able to get my head around what was going on.

I had the Diploma, so at least I knew the nuts and bolts but I took the view that the factory knew more about what was going on than I did, so I probably spent most of my time on the factory floor, drinking cups of tea, watching the guys doing the job and learnt about the job really from them, you know. Not learning about how to be a practitioner but the practicalities of putting in place controls and that sort of thing.

There were "brilliant" things that I tried . . . they didn't work. You know, I can remember designing elaborate near miss forms and they were just total rubbish. They stayed in the boxes, you know, didn't work. So I designed something simple. You learn from your mistakes.
Health and Safety Manager, Media Organisation

Of course, many practitioners will have developed their interpersonal skills in their previous work roles, and may be attracted to a career in safety because it affords them the opportunity to exercise some of the more subtle persuasive and problem-solving skills they have acquired. However, many have not acquired those skills. The behaviours which "worked" in previous roles may be completely inappropriate when a person moves into a safety role, or changes sectors, or even changes safety role within the same company.

[He] came through the construction route, and again his approach was "I just want to go and tell them what to do, and I expect them to

> *do it." But when he got a slightly more senior role, he was completely unable to do it, because he went in, tried the same approach, wound up the plant managers, and the plant managers, basically, locked him off site . . . And in that case he couldn't adapt his style, even with coaching. So ultimately it came down to, he's going to leave the business. And in the [dismissal] meeting when he was complaining that he couldn't get on the site, [so] he's being unfairly treated, he said: "Look, I just want to go out there and ******* tell them what to do" – in those words – "that's the job I want." Except he wanted like fifty grand (£50,000) for it as opposed to twenty-five.*
>
> Group Head of Health and Safety, Utilities Sector

There has been a tradition of those leaving the armed forces entering safety as a career. Many ex-service personnel go on to have distinguished careers in health and safety. However, a number do not, possibly because they anticipate that the role involves telling people to follow rules, rather than facilitating, influencing and persuading people to solve problems.

Given these challenges, with the positive impact that reflective learning and development can have on the effectiveness of safety practitioners, it could be argued that reflection should become an essential part of the HSE practitioner's education. As Schön put it "Enhancing the practitioner's ability for 'reflection in action' – that is learning by doing and developing the ability for continued learning and problem solving throughout the professional's career" – that is an essential part of the HSE practitioner's role. Using reflective and experiential learning not only improves the performance of individuals, it allows for a faster-track development of safety professionals and practitioners.

The effective practitioner

So what are the skills required to be an effective safety practitioner, outside of the technical skills which are likely to have been developed during a period of initial professional training? These will vary depending on your role, the company and the industry you work in and how senior your position in the organisation. Influencing, developing a business case and presentation skills are three common ones.

Influencing

Influencing skills are essential, but the need for them is more important in some contexts than others. There is a difference between a role based on one site, where the practitioner can more easily spend time on the factory floor and has frequent day-to-day contact with those whom he needs to influence, and may even be in a senior position relative to those people; and a "central" or group role, where results need to be achieved by education and influencing of stakeholders throughout the wider organisation with whom one may have infrequent face-to-face contact. In a single site organisation, the relationships that can be developed lay the foundation for influencing others.

Some individuals have very good influencing skills which seemingly come naturally to them and may be linked with the concept of emotional intelligence. However, many practitioners develop those skills through experience and reflection on that experience. Models of influencing skills can be taught and these skills then can be developed by practice. Numerous books have been written on this topic, but the key point is that to influence someone, you need to consider what type of influencing works for them, not what feels the most comfortable for you. Further, the need to maintain a long-term positive relationship with someone should inform your influencing style.[6]

One time, in [my previous company], I was presenting to some very senior managers. So I was trying to justify management involvement in some of the projects. My frame of mind was that these are senior managers, so I don't have to, shall we say, preach to the converted. Then one of the vice presidents said: "Management commitment? We gave you so many millions for your budget. Isn't that enough?"

This was a problem that needed management commitment, management concern, and their answer was, "We gave you a budget of, I think, about seven million." But it wasn't just about throwing money at the problem; it was about them walking the talk.

The lesson for me then was about buy-in, you know? I was under the impression that you don't need to preach to the converted, so all I needed to do was present to them what needs to be done and they will accept it and we will implement it. But there was no buy-in from the management. Without that, even with the best case made, it will not go through. Sometimes it also happens like, for example, there's

53

some professional jealousy. "I didn't think of this, you thought of it, so whatever the idea, I don't like it."

So now, I explain it to people individually first. I explain to them the rationale, the logic, the benefit and everything individually over a period of time, so when I finally present this to the senior management, I have friends in the audience who will say "Yes, that's good, we will go with it." But if you just present it outright without buy-in . . . it will bounce.

HSE Manager, Oil and Gas Sector

Business skills

It is often proposed that more business and management training is needed for HSE practitioners, so that they can ground their advice in business practicalities. However, reflective learning facilitates the development of the professional in the context of their day-to-day experiences. Reflecting on your experiences helps you to understand a situation from a number of points of view, including those of other people in your organisation. Understanding what motivates their actions at work is key to persuading them, and developing your own understanding of how safety, health and environmental management can help prevent loss in organisations. For example, a logical finance director could be approached with a business case showing how much absence is costing the organisation and the likely costs and savings associated with a programme to reduce absence. However, to do this effectively, you may need to reflect on and develop your skills in setting out a business case.

Communication and presentation skills

Making a presentation can strike terror into the heart of many practitioners and indeed many managers. And yet the requirement to deliver training, team briefings and make presentations on performance and plans is an essential part of most safety practitioners' job. As you become more senior, the need to deliver effective presentations increases, with the need to brief senior management.

In larger organisations, internal conferences presented to an audience of hundreds of safety practitioners, safety representatives, managers and contractors may be a real and paralysingly appalling prospect. Delivering an effective conference paper requires a slightly different approach.

Almost everyone at work will have had the misfortune to sit through a dreadful presentation, where perhaps the presenter was unclear themselves about what they were trying to convey, or where they attempted to pass on a huge amount of detail in a short space of time, or their delivery was "flat" and unenthusiastic or even worse, they had picked up someone else's PowerPoint presentation and were attempting to deliver it, having not even read through it beforehand.

There are a wide range of courses available on presentation skills. A good one will always deal with effective planning. Being very clear about the key messages you need your audience to understand by the end of your presentation is absolutely vital. If you cannot access a good course, find someone in your organisation or in your network of contacts who is brilliant at presentations, and ask them to coach you. Having completed the basic training, you need to practise, and you need to reflect on your "performance" each time. Smart phones and tablets now make it easy to have any presentation you give recorded. Use the recording to look at how you did and identify what went really well, what really worked and resolve to do more of that.

It takes courage to work on this skill if presentations are something that you are really uncomfortable with. But, like everything, practice will help.

Do we reflect in different ways?

One of the keys to improving your effectiveness as a professional is to understand yourself, and where your strengths and weaknesses lie. Understanding your preferred learning style may be a good place to start. The Honey and Mumford Learning Styles Questionnaire (LSQ)[7] can be completed online and can help you to identify your most and least preferred learning style (Table 5.1). That is not to say that you cannot learn in different ways, and indeed to learn effectively, you need to spend time in each stage of Kolb's learning cycle (see Chapter 2). Knowing your preferences can help you to devise learning activities that will work for you, and then you can work on the areas that will help you maximise that learning, in particular, reflection.

There is no one "right" or "wrong" way to learn. Some methods will feel more natural to you, and some you may have to grit your teeth to participate in. For example, if you are an analytical, logical Theorist, you are likely to find role-play of an inter-personnel conflict bewildering, particularly if you haven't

Table 5.1 Learning styles and preferences

Learning style	Description	Preferred ways of learning	Least preferred learning methods
Activist	Activists are those people who learn by doing. Activists involve themselves wholeheartedly in new experiences tending to be enthusiastic about trying anything new	• Group discussion • Puzzles • Competitions • Role-play • New experiences and challenges • Competitive teamwork and problem-solving • Practical work with an opportunity to "have a go"	• Passive methods, e.g. lectures or observing others • Analysis of data • Working alone, e.g. reading, distance learning • Dealing with theoretical concepts • Extensive repetition, such as learning a skill by practice • Following precise instructions
Pragmatist	Pragmatists need to be able to see how they can use their learning in the real world. Abstract theories and concepts have no meaning for them if they cannot see a way to translate them into action. They like to experiment with new techniques to see if they work in practice	• Case studies, preferably based on real life situations • Problem solving, preparing action plans • Discussions • Practical work with coaching or feedback	• Dealing with theoretical concepts which seem to have no practical application
Reflector	Reflectors like to think about experiences thoroughly, considering all angles and implications before coming to a conclusion	• Observing activities • Paired discussion • Self-analysis/personality questionnaires • Research • Writing reports • Coaching/feedback from others	• Situations where they must act instantly, without time to consider or plan • Where they are given insufficient information or time to come to a conclusion
Theorist	Theorists need to understand the theory behind their learning. They prefer logical systems and models, and to analyse their observations and new information in the context of their existing knowledge, synthesising the new learning into a "theory"	• In structured situations where they are intellectually challenged, for example, by analysing information and evaluating alternatives • Applying models, statistics, theories	• Role play and activities focusing on emotional aspects of a problem • Unstructured methods with ambiguity about what is required

Source: Adapted from Honey (1983).

been given clear guidelines in advance. On the other hand, an abstract model which fascinates the Theorist may be dull to the Activist, and of little interest to the Pragmatist unless he or she can see a way of putting it to practical use.

As we have discussed, Environmental, Health and Safety management is often a second or third career. It is likely that we choose our initial career based on a combination of aptitude, including preferred learning style, and opportunity. We may therefore bring those preferences with us into our safety career. All are useful, whether we are in search of the issues and challenges as an Activist, reviewing and reflecting on the experience as a Reflector, concluding from experience as a Theorist or planning for improvements as a Pragmatist. In all cases, reflective and experiential learning help a practitioner to discover more about themselves. This may in turn determine which skills we need to develop to be successful in our HSE job roles.

Identifying your learning needs

In their formal training, the would-be HSE practitioner learns that safety issues can be resolved by identification of hazards, evaluation of the risk involved and the application of a solution based on a set of principles which is commensurate with the risk and deals with the root cause of the problem. This is the "technical rationality" so eloquently described by Schön (1987) and described in Chapter 2. It equips the practitioner to deal with straightforward problems.

In their professional practice, HSE professionals more usually find that they are dealing with complex and unpredictable problems, sometimes these are beyond their "competence". Often a situation has implications which are beyond the formal responsibility of the HSE practitioner to resolve. For example, the solution may require a change in practice within the organisation or a department. It may require a pause in a process while additional tests are conducted or until a solution is implemented. In seeking to bring about an improvement, for example, preventing a repeat or a foreseeable future accident, the HSE practitioner often needs someone else to implement a solution. This can be difficult to achieve in an organisation where the HSE practitioner is in a relatively junior position and/or where there are no shared values on the worth of health and safety. Furthermore, a failure to persuade more senior people to implement a solution can present an ethical dilemma. Does the safety practitioner walk away, having given his best advice, leaving a hazardous situation? Or does he "kick up a fuss"? The first option may be contrary to his personal values and his professional body's code of ethics. The second may be career-limiting, and if done in the wrong way can undermine his professional credibility and jeopardise long-term relationships, making him less persuasive and effective in the future.

However, painful as incidents like this can be, they can also provide the most valuable learning opportunities, provided that the experience is reflected upon, and potential new methods of dealing with the situation are formulated, and tested out.

The critical incident as a learning opportunity

In our research for this book we spoke to a number of practitioners about their most difficult or painful situation they had faced at work. We are grateful to them for sharing experiences that often did not show them in the best light. However, all were able to highlight significant learning points.

I was escorted off site by security . . . I went with a colleague to work as an audit team, and we were there for three or four days, and I think we broke the cardinal rule of auditing in hindsight. So we did the opening meeting and we said: "This is what we're going to do, these are the risks that we think we're going to be focusing on, we're going to have a look at the management system then test it to see if it works, and we'd like to report back to you on Friday." And within a day we'd found quite a lot of what you might call findings . . . there wasn't enough water pressure in the sprinklers if they had a fire, and there was insufficient water to deal with it. I remember going to see the manager and I was told, "She's not here, she's too busy."

Over the middle days we just could not get to see the manager. Now, with hindsight, I would definitely have written an email or something to make her aware of what these issues were but, you know, hindsight is twenty–twenty. We got to the closing meeting on the final day and standing in her shoes and looking back, in front of her top team, she was embarrassed about things that she didn't know about, you know, and what do people do? They fight back . . .

. . . She was adamant that that the sprinklers worked so we ran another full flow test. I said: "Well, look, that needle needs to be on four bar and it's running at about 2 and a half . . ." We went back for finding two and she said: "If you'd been to see me, I would have told

you where the records were", so we marched somewhere else to find them and of course they weren't there and after about an hour of this, she . . . picked up the telephone. She said: "Hi, it's and her name, um, will you please make sure that [the auditors] leave the site?" and we were escorted off site . . .

But the learning was . . . never march into [a manager's] office and say, in front of [their] team: "There's all these problems." You need to talk to [the manager] in advance.

Health, Safety and Environmental Consultant, Insurance Company

I remember, as a very very young safety consultant, and I was probably about 26 at the time, not being paid. I wasn't self-employed; I worked for a company but basically I went to visit a site which was being run by a Major. He was no longer in the army but he was always called Major and I don't think he was terribly pleased that a young woman turned up to do his audit, to be honest. Basically he didn't pay because he said that I completely misinterpreted everything that had been said to me, which I hadn't . . . I think it was about criticising his actual approach as well as the kind of factual things he did, so, for example, accident investigation; the first question on their accident investigation form was "Who was to blame . . .?" which is absolutely not how we do things, but in retrospect a more tactful approach from me would have helped.

Health and Safety Consultancy Manager

In both of the above examples, it was not the auditor's technical skills that were at fault, though both were subsequently criticised by the client for that reason. The problem was in their "softer skills" in managing the relationship.

In both cases, there was reflection on the events and the determination to do something different. It is worth reviewing Honey's model of single loop, and double loop learning at this point (Figure 5.1), and see Chapter 3.

In single loop learning, we review what we did in light of the results we have received. We may conclude that we needed to better execute what we are attempting. In the example above, of the auditors who were escorted off site; the lead auditor might conclude that he needs to explain

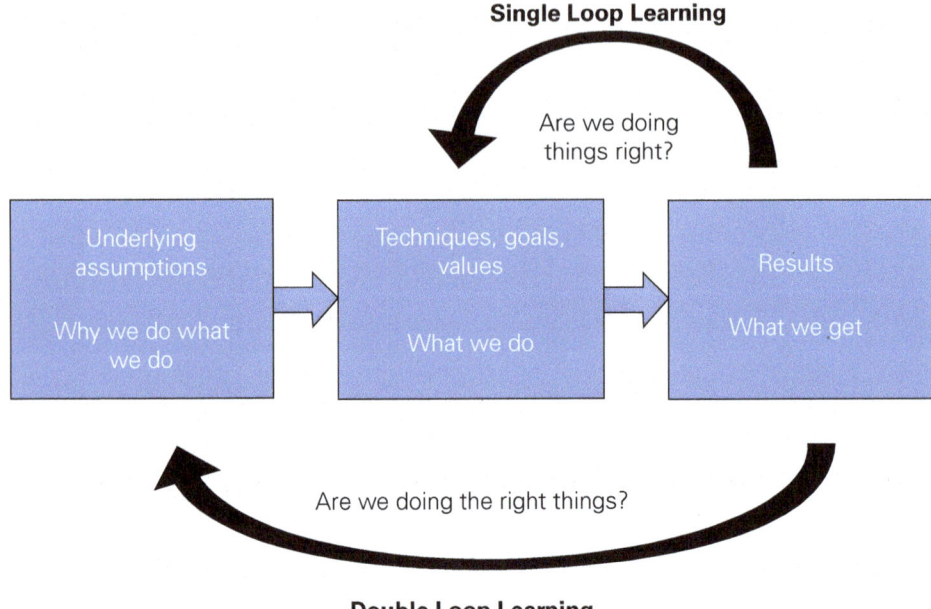

Figure 5.1 Single and double loop learning
Source: Argyris and Schön (1996)

the findings more fully, or more forcefully in order to make the manager understand that there was genuinely a problem.

In double loop learning, we go back to our underlying assumptions and question them. This can assist you to "re-frame" the problem. For example, concluding that you need to consider and re-evaluate your approach rather than trying harder at your original response.

There are neurological reasons why painful lessons tend to stick with us. Emotions are regulated in our brain by the limbic system, which plays an important role in processing memories. Memories formed during a particular emotional state tend to be easy to recall when we enter that state again. The limbic system and in particular the hypothalamus are responsible for the "fight or flight" response. Unfortunately this means that at moments of heightened emotions, we are at our least rational.[8] Reflecting after the event at a time when we are not so emotional can lead us to formulate a different response next time, provided that we can manage our emotions in the moment.

Often people describe a "gut feeling" that tells them that something is not quite right. (In actual fact there are neurons in parts of the body other than the brain, including around the gut.) Sometimes it takes a period of reflection to understand what our gut is trying to tell us. Perhaps we know that we have made an error, or we have behaved in a way which is contrary to our professed values, or we have an ethical dilemma. Our emotions are telling us that this is something that needs to be paid attention to and dealt with.

And one of the things we knew we had to change at [the company] was this "accidents are good because we get money for holidays", so we had to defend some of them. Quite often we could provide compelling evidence that at the time nothing was wrong. We once went to court and it was my call . . . and it was a guy who said he had been injured on a piece of machinery. And I can remember, we got all this stuff; we'd got a risk assessment, we'd got a then brand new certificate of conformity from the manufacturer, commissioning paperwork, we'd got our checklists, we got his training records, incident investigations; the stuff that a good defence is made of, and for about a day and a half I sat in the back of the court watching our managers go on and explain all this, this stuff and I thought, we're looking pretty good here. And on the afternoon of the second day, the plaintiff walked in and he spoke four words: "It wasn't that machine."

Well, it's powerful stuff because the thing is, you have to investigate the incident that's occurred, not the one that you sit in your office and someone tells you has occurred, so the immediate change, you know, eight o'clock the following morning was, "You will attach a photograph to your investigation" and I remember being asked: "What should the photo be of?" And I said: "You know, I don't care. I just want to know you've been there."

Health and Safety Manager, Engineering Company

Reflection as part of continuous professional development

Almost all professional bodies require their members to undertake continuous or continued (different terms are used in different member organisations) professional development – or CPD – to maintain their right to practise and remain a member. In some professional bodies, the CPD requirement is for a number of hours of learning or for the accrual of a number of "points" in a set period.

Some seminars or courses are advertised as being worth a specific number of "CPD Points" for a specified professional membership body. The huge drawback of this prescriptive approach is that not everyone will gain the same amount of development from attending the identical course. For example, if you are an experienced senior practitioner and you work in an organisation with a large number of employees undertaking industrial or construction activities, you are probably going to have investigated a significant number of accidents, including some serious ones. A two-day training course on accident investigation may teach you nothing at all. You may end the training thinking, "I could have delivered that course myself; I am more experienced than the tutor."

In contrast, a safety practitioner working in a smaller organisation in the service sector, even with the same length of service as the practitioner described above, may have investigated only two or three accidents and might find the same training course hugely informative.

Continuing with this example, however informative the delegates found it, the reason for attending the course is to improve their professional practice. That means that they need to *do something differently* because of attending the course. Whether they do or don't depends on a number of factors.

If the information presented on the course accords with what has previously been learned, or extends it, the participant may be able to draw connections, which will enhance the memory of what has been newly learned. It can be really helpful to physically draw this out, perhaps in the form of a mind map or diagram to show how the new information links with your existing knowledge.

If the topic is difficult to grasp, or challenges or is at odds with what you thought you knew, embrace this as a positive. If you have to really wrestle with a topic and ask lots of questions in order to understand, that usually means that you will form neural connections and lay down stronger memories of it. In the experience of the authors, we have found that those safety-related topics that we struggled with in our professional training actually became those we became the most effective at teaching as we finally grasped how the issues fitted together with what we already knew. Of course, if a topic is difficult, you have two immediate choices while on the training programme. First, you may choose to engage with it and try hard to understand, asking questions, talking to fellow participants, asking the tutor to explain it to you in a different way, or reading around the topic to see if that can help you to "get it". Or you can disengage, and spend the course mentally elsewhere. The first approach can take some courage because many people do not like to admit in front of others that they don't understand something, but ultimately it is the most worthwhile.

> *I left school at sort of 14. And I actually took the law paper four times but I got it. And when I got it, I literally danced round the room.*
>
> **Health and Safety Manager, Local Government**

Most important of all though, is to plan how you will use the learning and then actually use it in practice. This way, you get the most return on your training investment.

All of these factors – the existing knowledge and experience of the participant, their willingness and motivation to engage with the learning process and their intention to explore how to put it into practice – will affect the degree of learning which takes place, and the value that the experience has for them. In fact, what is being described here is Kolb's learning cycle which we discussed in Chapter 2.

The IOSH requires members undertaking CPD to write a reflective statement on the CPD activity and allocate "reflection credits". This means the member may reflect on the value that the learning had for them personally and give points accordingly. IOSH suggest that the reflective statement account should consider:

- your professional objectives in undertaking the activity;
- your approach and the reasons for it;
- details of your completed activity, including, where relevant, the contribution of others;
- the results of the activities and the extent to which your objectives were achieved;
- an analysis of the strengths and weaknesses of the approach you took, and learning points for the future;
- an explanation of how it may, or how you think it may, impact upon your future role as a professional health and safety practitioner.

A more developed approach, based on the theories of neurological learning that we have already considered and in particular the work of Moon,[9] might include the following components in a reflective statement or account (Table 5.2). The key addition here is that we consider our feelings or emotions around the activity which may influence what we choose to do next.

Table 5.2 Components of a reflective account

The **context** and **background** of the event/activity (e.g. I was participating in an accident investigation)
The **objectives** in undertaking the activity (e.g. attending a training course to understand the hazards of entry into confined spaces and the operation of this organisation's permit to work system)
Your approach and the reasons for it (e.g. I decided to use fault tree analysis methodology to try and better determine the root causes of the accident I was investigating)
What happened – an objective description of what was read, done, seen, etc.
Interpretation of what happened, such as main learning points, whether your objectives were achieved, how it fits with previous learning, why events unfolded as they did
Your feelings about the event, what was challenging or demanding, what you were thinking at the time and what you are thinking or feeling as you write the journal entry (i.e. as you reflect on the event)
The **strengths** and **weaknesses** of the approach you took, and learning points for the future (e.g. I learn best where there is an opportunity for interaction and questions)
Explain how it may **impact upon your professional practice** as a HSE practitioner. (What will I do differently next time? What approaches might I try next time? How can I put this into practice in my organisation? What extra learning do I need?)

Source: Adapted from Moon (2009).

Developing into a manager

Reflective practice is increasingly the subject for study within business schools. It is possible to undertake a Postgraduate Diploma in Reflective Management Practice. There have also been examples of experiential learning being the underlying ethos in some business schools.

[My MBA] was quite literally a life-changing experience. The ethos of the whole course was learning from experience, so they basically wanted you to work in teams for pretty much the whole thing, which was, you know, almost faking the experience of working in an organisation or in a project team. We didn't have many lectures, and, to be honest, when we did, we mostly found it quite dull, because we were all quite hands-on types. We explored the topics by doing projects, so, for example, we had to do a project on setting up a company, which meant doing the work on the market needs and how you'd position it, how you'd get finance, your initial business plan, and so on. It was actually great and at least two of the people in my year group left their jobs to set up their own business.

Health and Safety Consultancy Manager

In Chapter 4, we looked at some opportunities for professional development outside of formal training courses. Opportunities to develop your management skills can arise in similar ways.

One of my spare time interests is I play in a band, and have done for years, and if you can control a crowd in the Irish club on a Friday night, you can chair a meeting and I think that's where I got my baptism of fire, if you like. That's where I got my ability to stand up in front of people and talk and control people.

Health and Safety Manager, Local Government

Often if you are promoted into a managerial position, you will be given formal training to assist you to take on your new responsibilities. Many practitioners find that their training in safety is an excellent preparation for being a manager, and the systematic approaches taken to managing risk can be applied to many other issues. Skills of communication and influencing are of course relevant to all managers. However, a key aspect of being a manager is having responsibility for other people, and in particular being responsible for their development and performance. You should reflect on your own learning needs to undertake this onerous task. Examples of learning that you may find helpful are coaching skills, giving feedback, and performance management.

Creating the environment for your staff, team and organisation to learn

Managing other people is one of the most challenging jobs most of us will ever do, and not only do we need to acquire and develop our skills to manage others, we will also be responsible to some extent for the development of our team members.

We have an enormous responsibility and also an enormous opportunity to encourage our team to continue to grow and develop. Of course, training courses and the sensible allocation of the training budget play a part in that, but, as we have seen, there are numerous opportunities for the individual to forward their learning by other means. As a manager, we can create the environment and the opportunities for others to continue learning from their experience.

Most organisations have a formal annual appraisal in which the development needs of the individual can be formally discussed with their line manager.

Opportunities for secondments, project work, conversations and coaching with others inside and outside of the organisation can all be brought into the individual's personal development plan.

The opportunity to reflect is also very important. As a manager, you can encourage reflective learning in your team in a number of ways:

- Arranging a one-to-one meeting after your team member has attended a formal training course to discuss what they learned and how they are going to use it (a reflective conversation).
- Encourage the individual to develop an action plan with time-bound objectives, and review their progress at intervals.
- Ask the team member to make a short presentation on their key learning from an event to the rest of the team to share the learning.
- After an event or activity (not necessarily a formal training course, it could be a project, an investigation or an audit), hold a team meeting to formally review what went well. Dwelling on mistakes may be counterproductive in this group context, and inhibit open reflection.
- Allow your team to be formally mentored by others in the organisation and in turn allow them to become mentors. It will help develop their skills and challenge their thinking.

What I do now, if any of my team go on a training course, is sit down with them afterwards and go through what they learned. Then we talk about what they're going to do with it. So it's written into their plan. So [a colleague] went on this presentation skills course and first I asked her what the key things that had come out of it for her were . . . what had she learned? Then I asked how and when she was going to apply them. So we identified the next two presentations she was going to do and what she was going to do differently. We went through her materials beforehand. I sat through one because it was in a meeting I was at . . . that went well, actually.

It's a bit more difficult with my staff based outside of the UK, but I would still do it over the phone.

This is the crucial thing for me; it's important that when things do go well, to reinforce it by reminding them that it's because they have changed their behaviour.

HSE Director EMEA, Speciality Chemicals Company

Modelling behaviour

In conclusion, becoming a reflective practitioner is important for your development. It doesn't stop when you become a manager or a senior safety practitioner. It has to continue through your career. Modelling this behaviour yourself; showing your willingness to learn, to adapt your approach, to try new ideas out in practice, and maximise your learning from every experience, will help those you manage to understand the value of this approach and develop themselves to be better safety practitioners.

In brief

- ▶ HSE is most often a second or third career for practitioners. To retain credibility, the HSE practitioner needs to learn quickly.
- ▶ The skills that the practitioner needs to develop are complex and require significant interpersonal and change management skills as well as technical problem solving skills.
- ▶ Using reflective and experiential learning not only improves the performance of individuals, it allows for a faster-track development of safety professionals and practitioners.
- ▶ Reflect on "critical incidents" to help you to improve your performance by trying different approaches.
- ▶ Create an environment for others to use reflective learning to maximise the learning they acquire from their experiences as well as from formal training interventions.

Notes

1. Syed, M. *Bounce: The Myth of Talent and the Power of Practice* (London: Fourth Estate, 2011).
2. Marcus, G. *Guitar Zero: The Science of Becoming Musical at Any Age* (Harmondsworth: Penguin, 2012).
3. Gladwell, M. *Outliers: The Story of Success* (Harmondsworth: Penguin, 2009).
4. Anders Ericsson, K., Krampe, R.T., and Tesch-Romer, C. "The Role of Deliberate Practice in the Acquisition of Expert Performance", *Psychological Review*, 100(3) (1993): 363–406.
5. Reason, J. *Human Error* (Cambridge: Cambridge University Press, 1990).
6. Cialdini, R.T. *Influence: The Psychology of Persuasion* (New York: HarperBusiness, 1993; rev. edn, 1 Feb 2007).
7. See http://www.peterhoney.com/ (accessed 14 October 2013).

8. Peters, K. "Neuroscience, Learning and Change 360°", *The Ashridge Journal: Perspectives*, Spring (2011).

9. Moon, J. *A Handbook of Reflective and Experiential Learning: Theory and Practice* (London: RoutledgeFalmer, 2009).

CHAPTER 6

Methods of reflective learning

In previous chapters we have dealt with the theory behind reflective practice and reflective learning, explained the benefits to the HSE practitioner at various stages of a career and given examples of how HSE practitioners have used reflective practice to enhance their effectiveness. In this chapter, we will review some of the tools to assist with the reflective process and some of the factors in the modern working environment which influence both our need and our ability to reflect.

There are various methods of reflecting. These include tools for a person reflecting on their own and other methods which involve two or more persons in a reflective dialogue or discussion. As will be explained, no one method can be said to be "the best" as this is highly dependent on the practitioner using the tools, his or her preferences, the learning situation and resources. However, the strengths and limitations of each tool will be considered along with practical advice on using them.

Influences on reflective practice

There is a sizable amount of literature from the early 1980s onwards regarding reflective learning and also a great deal of agreement between academics that reflection on practice is important for the development of professionals.

The professional demands on the HSE practitioner have changed considerably over time and so have the learning methods and tools available. Some of the influences on the practice of HSE and the learning needs of the practitioner, as well as conversely some of the means of meeting those learning needs, have arisen from the influence of globalisation and working with multinational cultures; the influence of more regulations, standards and best practice; information technology and the internet; physical and virtual forums and fast-changing learning environments.

Globalisation

The impact of globalisation is that the geographical footprint of a corporation can span all time zones. Individuals working for such a business or with global clients may not find it unusual to wake at 4 a.m. to participate in a teleconference discussion on Skype. The HSE practitioner may have to respond to an urgent need for their expertise following an incident on the other side of the world, late into the night.

This tendency to demand an instant response 24 hours a day means generally that a person is more focused on the issues at hand rather than taking the time to reflect on what has occurred during that day or that week if it has been consumed by an important assignment. (HR professionals acknowledge that in some parts of the world, January is the best time to start a recruitment campaign as people taking time off over the Christmas and New Year break have the time to reflect on their work life, and conclude it is time to seek new opportunities.)

Of course, working in a global context also provides enormous opportunities to be exposed to different ideas, knowledge and experience. However, practitioners still need to take that time for reflection to assimilate their exposure to new information, ideas and experiences into their professional practice.

Working with multinational cultures

Perhaps in a way connected to globalisation, we find that HSE practitioners increasingly are having to work with multinational colleagues, contractors and clients. The differences may be less obvious than language. (It is often said that Britain and America are "two nations divided by a common language".) Work by Hofstede[1] and Trompenaars[2] helps to describe the differences between national cultures across a number of dimensions.

Understanding those differences perhaps helps us to realise that it is the interpretations we place on others' behaviour that is the problem.

Cultural differences can extend to our approach to health and safety. American standards are generally more prescriptive or rule-based in their nature whereas European standards are more goal or risk-based. (Even within Europe there are differences in approach. In the UK, there are standards in law which apply "so far as is reasonably practicable" which was challenged within the European Union as recently as 2007.)

In recent years there has been a move towards global harmonisation of many of the standards used in HSE, promulgated by the United Nations or the International Labour Organisation. However, even in the absence of harmonised standards, multinational corporations often choose to apply a common standard across their organisation. The standard applied can depend on the cultural allegiance of the organisation's head office which can prove challenging for HSE practitioners from a different background.

Let us take an example of the NFPA 30,[3] a US standard which deals with the storage of flammable liquids. In one clause, the requirement for intermediate bund walls between product tanks is required or highly recommended by the standard after the tank's capacity increases beyond 750 m^3. In a discussion with the project owner and the engineers, a senior safety practitioner reviewing the design of a tank farm containing eight tanks, each with a capacity of 730 m^3, insisted on intermediate bund walls to reduce risk. However, the engineers rejected it, based on the standard and the project owner rejected it based on the cost implications. In this case, a great deal of persuasion may be required to convince both parties to explore the spirit of the standard and to understand the risks involved and accept that intermediate bund walls significantly reduce the risk in the event of a leak from one of the tanks.

Some cultures are more compliance-based, taking greater comfort in sticking to a standard or rule, rather than looking at the spirit of a standard or exploring a risk-based solution which in the example described above would be going actually beyond code compliance, leading to impact on project schedule and cost.

Influence of regulations, standards and best practice

Codes and standards may not be strictly enforceable legal standards, but they have been developed in order to identify reliable methods for controlling specific risks. The existence of a code of practice or a standard tends to bring about standardisation as they are adopted by organisations keen to demonstrate good risk management, and standard working practices across their global footprint.

In many cases especially with respect to the British Standards Institute (BSI), the American Petroleum Institute (API), the UK-based Institute of Petroleum (IP), the US-based National Fire Protection Association (NFPA), the American Society of Mechanical Engineers (ASME), and the ASTM (formerly known as the American Society for Testing and Materials), codes are developed based on tried engineering practices and experience. They are a product of experiential learning, arising from intellectual and technical debates on the shared knowledge and experiences of specialists. It could be argued that at an early stage of this process of forming technical panels, someone has reflected that this is the best way to put a standard together.

However, the word of caution here is that standards and codes should not impede the process of reflection on the hazards and the best methods of controlling them in your particular workplace. After all, standards are developed in the context of a time and knowledge obtained through past experiences. Newer experiences, especially after a serious incident, may have a profound impact on future revisions of standards. Standards indeed may not represent the *best* method of controlling a hazard, but rather what is *practicable to achieve* in most workplaces.

In the UK, safety legislation is framed in terms of the duty holder doing what is "reasonably practicable". The legal interpretation of this phrase is that you must consider the risks involved in what you are doing, and what is both practicable to do to mitigate them (i.e. what is the best technology available to control those hazards) and what is reasonable to do in light of the risk. Risk is about the probability of harm arising and how severe it would be. In determining what is reasonable, the cost of control measures can be taken into account compared to the size of the risk. Hence, UK law does not require an over-reaction to a trivial risk.

More recent legislation may be considered to elaborate on this duty to safeguard so far as is reasonably practicable, by imposing an explicit duty to conduct and review risk assessments for the purpose of determining the best means of control. In fact, this goes beyond simply the technical risk controls, it is about having the measures to implement the technical controls consistently. The ethos of the UK legal framework implies a requirement to reflect on what you are doing, and to determine if your risk control measures remain appropriate, particularly if the risk changes. The probability of doing something wrong can increase in many circumstances, for example, changing a contractor, carrying out a task in extremes of weather or under time pressure.

Even if you are operating as a HSE practitioner far beyond the UK regulatory regime, the authors strongly suggest that practitioners and the engineers, designers and scientists they work with, should continually reflect through challenging the standards that they work to and implement every day. This is a dynamic reflection to determine if those standards are still the best fit for their working environments. After all, this is one of the logical ways to arrive at what may be termed "best practice".

Information technology

Information technology is a double-edged sword. It provides fast access to information and knowledge, yet the great temptation for a HSE practitioner to "copy-and-paste" information available on the internet may impede reflection on that information.

In researching this book, it was possible to identify material of interest quickly, access it easily (not always free of charge) and hence be able to compare, evaluate and reflect on its implications for the HSE practitioner. In this sense, it was not unlike conducting research 30 years ago, except it was quicker and could be done from the office, at home, on trains, in hotel rooms and even on an international flight. This allowed the book to be completed in a shorter time period, but it did not save time in the actual writing, which very much involved reflecting on the research, formulating our thoughts, preparing drafts of the chapters and reflecting together as the content of the book emerged.

HSE practitioners with a particular technical problem at work can access a plethora of international standards, and codes of practice. They can have access to technical reports on major incidents as soon as they are available. These are all positives.

However, one negative consequence is that the ability to "cut and paste" whole chunks of text instead of the traditional approach of reading it, evaluating it and applying it to your problem or situation comes at the expense of engaging with the information, reflecting on it and formulating new ideas. Academic assessment has had to find ways of identifying where students have plagiarised work from the internet, and has applied appropriate penalties, such as disqualifying students from their degree.

Linked to this is the expectation of a speedy response because "everything you need to know is there on the internet". This has undoubtedly put additional pressure on practitioners to meet a performance expectation to produce and disseminate information quickly. However, this may have the tendency to inhibit your critical reflection on the material you uncover. Not everything on the internet is fact. Decisions made without appropriate reflection may be bad decisions, and the absence of reflection means that learning from the experience or the problem to refine your professional practice is absent.

We have an on-line accident reporting system at work, and in some ways it's great. You have automated workflows so that the right information goes to the right people very quickly. The challenge is that you have no filter on what's going in. So you obviously want the incident input to the system as soon as possible, but at that time the line manager may not have all the information or their report might not be accurate. Then you have head office clamouring for information you've had no chance to collect.

In fact, sometimes it's a barrier to good analysis of accidents. You have a pick list of potential causes and I wonder whether that triggers the choices people make. Then when you come to run a report off the system, people may ask for immediate causes and conclude, "Oh problems with PPE are causing a lot of our accidents" whereas if you dig deeper, there are more important underlying causes.

It should help you to focus your efforts, but if the accident isn't categorised properly or input properly, or the person running the report hasn't been trained, then what comes out is just misleading.

European HSE Manager, Engineering Sector

Physical and virtual forums

Direct communication between HSE practitioners has been enhanced greatly with the availability of both physical and virtual forums. Practitioners can meet each other in person at meetings of their professional body, both nationally and regionally at "branch" or "chapter" meetings. Some professional bodies have specialist interest groups for members working in a particular industry. There are also industry forums, trade associations, government-led discussion panels, conferences and workshops that practitioners may attend. This networking has facilitated that transfer of knowledge, experiences, and best industry practices and greatly helped in mentoring and developing new entrants to the profession. What it has also done is provide an excellent opportunity for reflecting in groups.

The same can be said about virtual forums where social networking sites host professional groups, for example, on Linkedin. These sites allow interest groups to share information and experiences. In a more structured learning process, webinars have also become common and people frequently use these to learn about new things and develop new knowledge and skills. Reflecting in such forums is also made possible with many individuals sharing their experiences or thoughts, and some share the lessons learnt on a personal basis, which helps others gain some insights into both the common and differing challenges they may face on a day-to-day basis. Some of these forums are also facilitated virtually and a good facilitator of discussion can help encourage those participating to share their experiences and the insights gained through their reflections on those experiences.

Of course, there are also drawbacks. On some chat forums, the person sharing their experiences or views on resolving a problem may just be plain wrong. Often moderators do not intervene unless a posting is offensive or abusive, which means that incorrect advice remains in a public space. Critical reflection on the information you find on the internet is advised.

Reflective learning tools

When one of the authors was at university in the early 1980s, there was a discussion in a "complementary studies" session on the implications of information technology. It was a widely held belief at that time that the main change would be with the ability to rapidly share information, and complete tasks more rapidly, that we would all have so much more leisure time. In fact, the opposite has occurred. With the ability to achieve more in a shorter time, the expectations of what can be delivered by the individual have increased. The HSE practitioner needs to learn more things and process information at a much faster rate. The downside of this has been diminishing time to reflect. The naturally calmer times of the day have receded in a 24/7 business world.

However, the HSE practitioner must make time to reflect. In the following section we consider the various methods and tools to assist you to become reflective and maximise your learning. It must be appreciated that while these tools do exist, it is impossible for us to confidently state which tool or method is the best. We can say that certain tools and methods will be easier for some to use than others. In certain contexts, some of these tools are well worth employing as part of a formal process, for example, reflecting within a project team with effective facilitation at close out may be extremely valuable for the development of all participants.

Your personality may determine which methods feel more natural to you. For example, introverts may find keeping a reflective journal a useful way of reflecting as an individual. The reflective journal can be written or recorded audibly. Either way, there must be some way to read or hear the thoughts expressed at the time of writing the journal and then reflect on your thoughts about the experience.

Extroverts may prefer reflecting with others in a group or in a one-to-one conversation. When reflecting with others, the process differs in that there is a sounding board. Extroverts who may be less in touch with their interior world may find it useful to have others reflect back to them an outsider's perceptions of the emotion that the extrovert is exhibiting in describing an event. It is a less introspective process than where you are effectively talking to yourself.

In this section we discuss each method and give some examples of where they can be effective.

Reflecting on your own

In Chapter 3, we briefly reviewed some of the reflective learning methods. One important point is that the methods are not mutually exclusive. For example, keeping a reflective journal which focuses on critical incident analysis combines two key ways of reflecting as an individual.

The reflective journal may superficially resemble a written structured "diary" of events. It could also be a recording either transcribed by the person him/herself, or by a transcriber. Some advanced voice-recognition software can transcribe quite effectively one's narration. Many individuals now have their diary on-line in a blog in which they record and share the events of their life. Some have a "vlog" or video diary, again which may or may not be available for public view.

Whether the log is written or a voice-recorded file, the ideal is to record your reflections at a peaceful time of the day where you are relaxed and undisturbed. It is recommended that critical events are logged down quickly as soon as possible after they happen, and then the reflection on that event can take place later. For the HSE practitioner, the triggering event could be a situation on site, an interaction, a meeting, a response to some communication or any other event, positive or negative, on which it is felt to be worthwhile reflecting. In Chapter 5 we looked at what you could include in a reflective account of an event that might be made in a reflective learning journal.

Returning to the concept of reflection briefly; a definition we considered in Chapter 3 was:

> being mindful of one self, either within or after an experience, as if a window through which the practitioner can view and focus self within the context of a particular experience, in order to confront, understand and move towards resolving contradiction between one's vision and actual practice.[4]

In this process the practitioner becomes more aware and gains a powerful introspective insight which leads to developing practical knowledge and wisdom when dealing with future situations. In that respect, it is useful to look back at your entries over a period of time. The two examples below, relating to similar events some months apart illustrate this.

Date	January 2010
Activity	Job Evaluation Committee – The evaluation of all the jobs within the Occupational Health Function.
Critical incident	Challenging the Committee on many of the technical jobs.

Reflective account

After a long wait, finally the Job Evaluation Panel has been convened. After a review and revamp with major changes in operations, I sent in 12 different jobs for re-evaluation last August. My employees are really unhappy that they have had to wait so long. I am glad it finally has happened.

The meeting or at least my presentation which was supposed to take 30 minutes, lasted nearly 3 hours, because the panel members, who obviously had not read all of the JDs or the presentation material beforehand, drove me up the wall asking me about the smallest details and arguing with me on why and how and really challenging the technicalities rather than challenging the actual job accountabilities, knowledge levels, etc. as they are supposed to.

Afterwards the Committee Chairman did pass on a remark to me that I was the first manager that had behaved in this way; who actually argued with the committee members forcefully rather than trying to keep them happy as they are the evaluators at the end of the day. I think my explanation back to him was that I believe in the process, and when I see that the meeting is really going in another direction, I feel I need to make my points more forcefully. I think the problem is not so much that things are not clear – it is that the members don't do their preparatory work which is what the process requires of them. This leads to the jobs presenter having to explain verbally what has already been submitted in writing. I think my challenge is greater because my function is so specialised and people just not understanding what we do. I really have to think of a way to communicate better or at least more effectively with them.

Date	May 2010
Activity	Job Evaluation Committee – The evaluation of all the jobs within the Environmental Function.
Critical incident	Challenging the Committee on many of the technical jobs.

Reflective account

Once again another round of punches to and fro. What was interesting in this meeting was the Committee Chairman's observation which was that I actually addressed the presentation more systematically and had already built up some good arguments about parity issues rather than what I did in the previous evaluation meeting. I appreciated that comment as I felt that at least my new strategy was better than the last time – although once again the 30-minute meeting still took about 1.5 hours when it really required about 20 minutes. I wanted to get through all the jobs but they were only willing to do the environmental positions. This annoyed me from the start of the meeting but I went on.

Part of my frustration is that I know in reality it has nothing to do with the committee and the evaluation, it is the process which is flawed right from the start. I know the system and the evaluation process. I have been trained, for God's sake. It was just very clear they needed to have these jobs properly screened by SME way before they were presented to a bunch of just "trained assessors". They keep on asking what does this mean, and what does that mean, and if this is really needed for the job! I felt that both my credibility as the functional specialist/head as well as

> my intelligence were being tried and tested again and again. If I did not think they did that or had to do that, I just would have not put it in the JD in the first bloody place!

As one can see, both accounts deal with an event which had both technical and administrative aspects to it. This is a frequently occurring scenario in most managers' job roles. What can be seen from the two accounts is that there has been a development in the approach from one incident to the next. Perhaps the challenges did not change but there has been a more positive approach in which some of the frustrations remain, but dealing with the realities more objectively helped the practitioner overcome the issues that occurred in the first incident.

It can be argued that while these logs are fictional, they are very much based on reality. A practitioner allowing themselves to reflect after critical incidents can help bring about an improvement through the application of the double-loop learning discussed earlier in this book.

But reflection can be used after positive incidents too. Reflecting on what went right sometimes is also an extremely valuable exercise. A case in point is this example.

Date	October 2011
Activity	Opening of the Occupational Health Services Centre.
Critical incident	Moving the Clinic from outside the premises into the Head Office premises.

Reflective account

Finally and after a tremendous 3 years and 10 months, I was able to do it. Set up the Occupational Health Services Centre within the heart of the Company's Head Office. I have finally done it with my team. It has not been easy to get all the approvals from management on the costs, the area and the idea of employing more nurses. The other approvals from the Government have been complicated as I think we have become one of the very few companies allowed to have our own clinic. Just going through and explaining to so many people the idea of setting up a preventative medicine outfit was a challenge. Indeed, it took time, but eventually through engaging different stakeholders, keeping the team appraised and motivated brought about success.

I can't remember a time that I was so happy and proud of what I have been able to do. To drive this through our organisation since I

> *started working on the initial research in 2003 has been great. I was unhappy really with the current operation, with our clinic running out of a partly rented clinic down the road. The clinic was small, so small it was claustrophobic! I didn't feel it gave the employees and the managers a good feel about it and made it more difficult to access preventive medicine services. I am optimistic that this new clinic is going to be a success.*

Here once again while there is only one log entry represented, the practitioner has reflected positively about what happened and how success was gained from effective stakeholder engagement. It may be noted the time frame for success was not short. We can read this log in a different, and perhaps more negative way. The process took too long and the benefits to the organisation were delayed. However, often as a practitioner we need to be patient where we are trying to convince non-specialists of the need for a change. Sometimes, as in the example described above, we need to take others on a journey with us so that they fully engage with the need to implement a very novel (to them) idea.

One might also think that this account could be further usefully developed by the practitioner identifying the key behaviours which led to this success, and the different approaches that worked with different types of stakeholder, but actually this reflection occurred throughout the long process of developing the case, winning support and working towards opening the new clinic.

Exercising patience and persevering through the process are part of the job of an HSE practitioner who is trying to bring about organisational changes. Re-reading a positive account of something that delivered success can buoy up your spirits as you work through your next difficult project.

Some tasks do not require the same amount of organisational change. However, it can be argued that HSE practitioners often need to work on bringing about conviction in others on the value of good HSE and securing their commitment through policy development and initial roll-out. Perhaps a greater challenge is maintaining the momentum of colleagues and keeping the organisation on track, post implementation, and into continual improvement. When we reflect as we go along, we are able to reinvigorate our ideas and strategies, learning from what went right as much as from the incidents where things did not go so well. In short, when we reflect as individuals we must believe that any and every incident is a great learning opportunity – by reflecting, we are able to gain from any experience. This clearly creates for us over time, accelerated experiential learning opportunities.

Reflecting alone does not suit everyone. Sometimes when we have fixed ideas about a situation or event; a "tight construction"; it can be difficult to consider it from other viewpoints, or even admit to ourselves that there may have been other or better ways of handling it. In these circumstances, discussing it with someone else, formally or informally, can be extremely beneficial to our reflection and learning from that event.

Reflecting with others

Working with a mentor

Many people find it easier to reflect with someone else. This may be informally talking over a situation with a friend or colleague. Working with a mentor is a more formal process and involves working with a more experienced, not necessarily a more knowledgeable, person. This is an important point. If you are a specialist within your organisation, there may not be anyone available to mentor you who knows as much about HSE as you do. However, as we have seen, many of the challenges involved in developing as an effective practitioner link to our emotional intelligence, our relationships with other people in the organisation and our ability to achieve results through influencing others. It is these experiences, if shared, which allow the mentor's wisdom to come through, to help the person being mentored to find their own solutions.

Different mentors will have differing approaches to the process. In general, good mentors will use the following:

- active listening to the person's experiences and stories;
- allowing time for the person to re-live the experience and describe it which in itself can help with reflection. This is a very therapeutic process for many of us; especially when recounting a situation which we found challenging, painful or we felt we were misunderstood;
- probing with questions to confirm understanding;
- showing empathy, especially when challenging experiences are being described. Good mentors are not judgemental even when they disagree with the actions described.

The role of the mentor is not to *tell* the person they are mentoring how to do something, but to help them to identify the meaning of the events they are describing, to consider and evaluate alternative approaches and to identify opportunities for trying a different approach, and then reflecting on how that experiment worked. Another way of looking at this is that the mentor is helping the person to progress through Kolb's learning cycle.

Sometimes mentors are able to assist the mentored with insights from their own working life but the focus must be on helping the mentored to reflect and build on their own personal experiences and here really is the benefit of reflecting with a mentor. The mentor can help the reflexivity process to take place and overcome the blockages which may exist when reflecting alone.

HSE practitioners who are fortunate enough to be part of a larger team may have formal and informal mentoring arrangements with other team members which may focus on technical aspects of the job. Below is a short account from a Safety Advisor who has been in the mentoring relationship from both sides.

Yes, we had to present a journal [as part of the requirements to be a registered safety practitioner in his home country] but what I gained from that actually is that I recognise that this profession requires a mentoring system. More than other professions actually, to be honest with you, because there is only so much you can learn from the books, from the classes, you know . . . It is this wisdom acquiring, you need to be guided. I have learned that you can have a lot of experience but what is more important is the correct *experience.*

Now I have an item on my job description to develop nationals [in the Middle East] but previously I have formally mentored three or four back home [who were becoming labour inspectors]. I have two pieces of advice I gave to them. One is that you have to be patient, you have to give yourself time to finish the course [because if you don't], I will never recognise you as competent. And number two is that if you choose to leave the organisation, if you go to another one, make sure that you have somebody you report to. Especially those who are just purely looking [at the salary]. Well, I told them: "Look, they may pay you maybe an extra two, three hundred bucks to go, but you're not going to learn."

You need them mentoring for the first three years. It depends, some people need a bit more or less. But the first three years is very critical. I think it gives you that maturity and above five years you need to find your own style and learn to do things your way.

Safety Advisor, Oil and Gas Sector

Mentors are sometimes trained in mentoring skills. Others may not be trained but have a personality that leads them to being naturally good mentors; as a good listener and an empathetic and non-judgemental person. More senior safety practitioners in that role can use reflective practice to improve their mentoring skills.

In my current situation, it tends to be more of a coaching-type approach. I have done a number of trainer training courses. But what I find is that you draw from other different inspirations. You know, for me, it could really be watching a TV programme and I wouldn't be really into cookery programmes but you could look in there and say, actually, I quite like the way that chef gave feedback to those people . . . And so you take that on board yourself so it's not just, it's not just formal training but also wider training, as I see it.

Safety Director, Global Facilities Company

One last very important aspect of reflecting with a mentor is that the practitioner must set aside the time, and make a commitment to respect someone else's time. So those practitioners, who are less self-disciplined at creating time and space for self-reflection, may find this process more suitable for them.

It is possible that practitioners may start reflecting with a mentor and later, once they see how the process is helping them, develop the skills to start reflecting as an individual using a reflective learning log or otherwise. Sometimes indeed the log or reflective account may be suggested by a mentor between meetings as the basis for discussion at the next meeting.

Working in reflective groups

Working in a reflective group is a process commonly employed by therapeutic support organisations such as Alcoholics Anonymous, or clubs supporting those trying to lose weight. There is no doubt that such groups can be effective in helping people to change their behaviour, but at the same time there may be a stigma attached to this approach. There is the implication that attendees have a problem.

Actually a reflective group, whether it is a long-standing group, or a short-term arrangement, for example, during a training course can lead to an exceptional learning, reflecting and self-development opportunity. It is similar to working with a mentor, in that an individual may recount an experience that they have just had, but instead of their dialogue being with just one person, they are sharing it with the others in the group, who may seek to clarify or challenge their interpretation of the meaning that has been placed on the event. At other times it may take the format of a group discussion on a topic.

Opportunities to form reflective groups can arise in a workplace where there is a group of practitioners, or in the context of a specialist or development group within a professional body. It can also be the mode of learning used on development courses or programmes.

More extroverted individuals may find this process is very suited to them. Introverts, on the other hand, may find it torture.

To gain the most out of group reflection, three things must be in place:

▶ a good facilitator, who can manage the discussion to keep it focused on learning, and ensure that everyone gets an opportunity to participate, share ideas and experiences;

▶ the group members are focused on learning (rather than on just expressing their opinion), and understand their role. Any feedback they give must be empathic and supportive so that participants are willing to share experiences, yet critical enough to add value in their feedback.

▶ a certain amount of trust borne from mutual respect. Some of the critical incidents which are shared may be very personal to people, either the experience in itself or the thoughts and feelings that followed. Often we need to learn from incidents which we did not handle well. We need to trust the people we are working with before we will disclose events which do not show us in our best light.

All participants can benefit, not just the one who is sharing their personal experience. You may gain insights into the experiences and challenges of other practitioners which may be similar to yours or may describe situations you will face in the future. Also, listening to various views and experiences may trigger self-reflection (which you may or not choose to share with the group), and help you gain insights even after the session has completed.

Basically a call came in to the office about a problem in one of the primary schools, and I was the only one in at the time, so I agreed to go. The Head was annoyed because the school had a gym lesson scheduled to take place in the school hall immediately after lunch, and the two catering assistants were required to move all of the lunch tables and chairs and stack them in an alcove before this lesson could start. He wanted them to do it quicker, and they said that they were being asked to work in a dangerous way. The situation had obviously been brewing for a few weeks.

I felt in a no-win situation because everything I said just seemed to make the Head angrier. For example, I asked if someone else could help them to stack the chairs so it could be done quicker, which I think he interpreted as me siding with the dinner ladies. And to be honest, mentally I was, because I thought he was being really unreasonable about it, and he was pretty aggressive and arrogant. In the end, I suggested to him that this wasn't really a health and safety issue. The manual handling wasn't a significant risk. So I left. Job done in my mind.

About an hour later, I'm back in the office and my boss receives a call and basically gets an earful from the Head, then puts the phone down and asks me what the hell I thought I was playing at. I was pretty upset and snapped back at him that he could have at least found out from me what had gone on before concluding that I'd messed up.

I got the opportunity to reflect on this at length about a week later. I had been booked onto a course about assertiveness and they invited us, in a group, to discuss situations where we felt we hadn't been assertive. I raised this one, and we ended up role playing the events. In the role play, I played myself first, and then the facilitator suggested I played my boss.

I realised as we did it, that actually I'd been a complete idiot, in all of the events of that day. I might have been technically correct in what I was saying, but my communication skills with the Head were non-existent; I'd let my views of his anger shape how I responded, and far from helping him solve his problem, I'd inflamed the situation. As for my boss, well, if I had told him as soon as I got back in to the office what had happened, then he could have dealt with a situation that I had totally failed to. I might have learned some better ways as a consequence.

So, yes, not the event, but the opportunity to reflect on it after really changed my approach . . . Also, if I hadn't been in that group at that time . . . well, I'm not saying I wouldn't have learned the lessons, but it would have taken longer to get there.

Safety Officer, Local Authority

Reflective practice in projects

Projects generally give an extremely powerful learning experience, as discussed in Chapter 4. There are many reasons for this:

▶ They are generally time-bound.
▶ They are generally well planned.
▶ They have key targets and milestones which can be used as checkpoints to stop and reflect.
▶ They involve many practitioners at one time.
▶ They involve a multi-disciplinary team of specialists and generalists.
▶ Key responsibilities and accountabilities are clear.
▶ Persons involved in the project team are focused on a common goal – to deliver as a team.

The process of reflection here is similar to reflecting in groups. Project-based reflection at project milestones or otherwise after critical incidents or even at the end of a project is important.

Getting used to working with each other in a group is usually a very big challenge in itself.[5] A description of the stages that groups go through before working effectively is described in a model by Tuckman, which is shown in Figure 6.1.

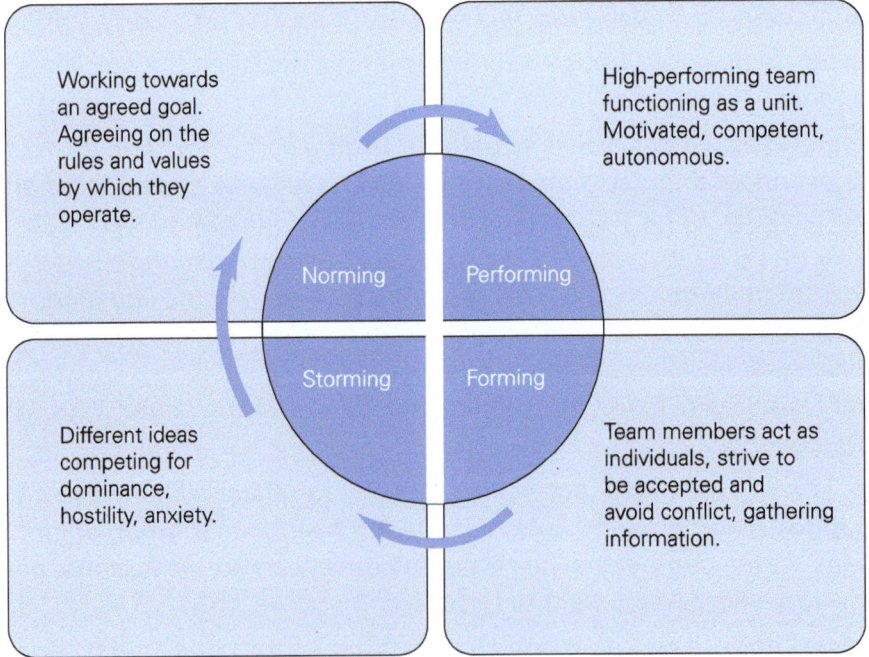

Working towards an agreed goal. Agreeing on the rules and values by which they operate.

High-performing team functioning as a unit. Motivated, competent, autonomous.

Norming   Performing

Storming   Forming

Different ideas competing for dominance, hostility, anxiety.

Team members act as individuals, strive to be accepted and avoid conflict, gathering information.

Figure 6.1 Stages in the formation of a team

Williams[6] conducted research on how project teams perform and his findings are illuminating in terms of reflection. He explained that the team members are generally rushed during and immediately after the project and find it difficult to take time out as a team to reflect, and he emphasised the need for management support to give time to the practice of reflection and learning from projects. Reflecting as a group, with the primary focus on "what have we learnt from what happened?" rather than "whose fault was it that things did not go to plan?" is pivotal to the success of learning for future projects.

This concept will be familiar to those HSE practitioners who have participated in effective team incident investigations which pull together the expertise from a number of individuals, working towards a group goal of trying to ascertain the root causes of incidents and the measures to prevent their reoccurrence. Incident investigation teams are effectively project teams – they discuss facts and try to ascertain what went wrong. They focus on human behaviour as well as management system failures and thus much of the learning comes through reflecting both as individuals and in a team. In fact, reflection within the incident investigation team can be wider than the actual incident being investigated, and can move on to consideration of other untoward events which the root causes of this incident have shown to be a possibility.

Project teams can always improve their performance through reflection. In fact, they have to, in order to improve. While many would argue that the critical factors for success of project teams include competence of the members, knowledge, management commitment, client understanding, good contractor management, and communications, reflecting on critical incidents as a group will generally improve performance both within the same project going forward as well as on other future projects.

The biggest challenge for a project team is time. The project manager's belief in setting time aside for the team to engage in reflective dialogue is fundamental. This might not be a long process, but there should be a "take-five" philosophy where focused, reflective discussion on lessons learned is provided for the team. Improvement can impact on all the aspects of the project, such as better planning, coordination, process improvements, and closer supervision of particular aspects. Projects are dynamic, time-bound, generally performance-based activities. Changes are a fact of life and therefore, taking time to stop and reflect and then brainstorm ideas is an invaluable process. Corrective actions made in good time save further potential delays, cost escalations or even injuries and incidents.

Strengths and weaknesses of these different methods and their application in practice

There are obvious benefits and strengths for each of the reflective processes described. There are also weaknesses or more accurately challenges to employing them. Table 6.1 summarises these.

Table 6.1 Key strengths and challenges of each type of reflective learning method

Method	Strength	Challenges
Reflective journals and critical incident analysis	• Readily available and free of charge • Flexible in level of detail and frequency of use. • Account is for personal use so user can be extremely frank • Creates a very personal learning experience • Can be therapeutic • Works well for more introverted and introspective practitioners	• Requires self-discipline to set aside the time • Less suitable for those who dislike the written form • Requires commitment to engage with reflecting on the material and the practitioner may get "stuck" in one mindset
Working with a mentor	• Very good for a practitioner to access wisdom from another experienced person/practitioner • Easier for many who do not want to use the reflective journals and Critical Incident analysis method or find it challenging to be disciplined with time • Excellent for developing managers	• Time must be set aside for the process • Dependent on availability of a suitable mentor • Requires commitment from both the mentor and the mentored • Effectiveness highly dependent on honesty and transparency • Competence of the mentor is critical
Working with reflective groups	• Excellent opportunity to meet with like-minded practitioners who may be facing similar challenges • Can be an extremely energising process • Practitioners potentially learn from other practitioners' experiences accelerating their own reflective learning • Encourages reflective learning which may help the practitioner learn to reflect on their own	• Requires that all the team members remain focused on the reflections and learning • Must have a good facilitator • Requires trust and respect from all members • Requires a discipline with time to have a group get together • Participants may not be in the right frame of mind to reflect at the scheduled time • Potentially more suitable for more extroverted individuals
Reflective project teams	• Contributes to continual improvements • Manages change while getting members involved in learning from the process of change management • Helps build strong teams with greater rapport • Creates learning that can improve the effectiveness of future projects	• Highly dependent on team members and their acceptance of this process • Requires supportive Project Manager • Requirement for project time which may be set aside if the project does not run to plan

What is important to appreciate here is that there is no one process which is better than the others. They all have their fair share of strengths and weaknesses, and many methods really incorporate elements of the others, or trigger greater effectiveness in the other methods. Thus the reflective HSE practitioner should explore all available methods, to gain the maximum learning advantages that each process provides.

Fundamentally, reflective learning is a very personal process which is dependent on the individuals taking the time and creating the space in which to reflect. Reflecting as an individual, with a mentor or in a group all require some degree of discipline. Only once a person is convinced that reflective learning can ultimately make them a more effective practitioner and help in their development will any of the tools be used.

Conclusion

This chapter has considered some of the challenges that practitioners face which impact on their ability to reflect. It gives a practical guide to four of the tools that can be used by the HSE practitioner. It is for the practitioner to really decide if they want to engage in reflective learning and, if so, which method to use and when, during their career.

The authors strongly encourage using a reflective journal particularly for Critical Incident analysis as this is a very powerful method. We have given in the last part of this chapter the strengths and challenges of the four main processes. We also, for completeness, present Table 6.2, summarising the application of reflective learning tools for the HSE practitioner.[7]

Table 6.2 Application of reflective learning tools for HSE practitioners

Reflective process, tool or method	Basic description of approach	Application in HSE management	Advantages and critique
Story-telling	Using a mentor or facilitator, the practitioner is encouraged to articulate the problems and events in their own narrative – meanings are then jointly explored	Narration and construction of events can contain critical incidents which can be studied – these are also relayed or written down from one's own perspective and using one's own language	This encourages a review of experiences and events and facilitates a good learning process through reflection with someone. This could be a mentor or senior practitioners who can comment and give examples from their own experience

Continued ▶

Reflective process, tool or method	Basic description of approach	Application in HSE management	Advantages and critique
Reflective and reflexive conversations	This is done in pairs and this is done frequently between practitioners – can be done with an experienced partner	Reflective conversations usually aim to discuss critical incidents. This is especially useful in working on strategies to address safety culture and development challenges	This aids reflection in action and the search for new meanings, options and perspectives
Reflective dialogue	This is generally done in groups where the facilitator encourages talking and also encourages the group to challenge some set norms and assumptions and encourages healthy and constructive debate	This is a method used in brainstorming in safety workshops and seminars. Working groups can work on debating policy issues, performance-based issues or even quite technical matters based on their working experiences and exposures	Very much dependent on the skills and experiences of the facilitator, size of the group and mix of people and experiences. A competent facilitator can encourage even the most introverted person to learn from observing the discussions and debates
Reflective journal and critical incident analysis	This is a highly individual approach in which a diary is maintained. It is more structured and calls for a person to log down experiences and critical incidents	Maintaining such a diary/log/journal helps note down all experiences as an opportunity for learning. Reviewing the journal over time can help the practitioner to evaluate different approaches they have tried and reflect on what went well as well as what could be improved	The learning takes place when incidents and log entries are analysed by the individual and categorised in terms of learning. They then also have a chance to see how better to approach the same issue if faced with it again. The main challenge is that at the time of the incident, this particular incident may not have been classified by the person as critical. Through reflection, such incidents can have a great impact on change
Concept mapping	This can be used by individuals or also used by groups. Here the facilitator or person explores how and why concepts are selected and linked	This is useful when we are looking at incident investigation and analysis	HSE practitioners today face many challenges which manifest themselves as safety issues, but in actual fact are related to other issues such as HR (e.g. staffing levels); design (e.g. substandard controls); funding (e.g. lack of attention to budget requests for upgrades/updates). Concept mapping can help identify where a problem lies

Source: Budworth and Al Hashemi (2012).

In brief

▶ The increasing demands of a globalised business world and the demand for speedy responses can make it difficult to find the time to reflect.

▶ The internet has helped HSE practitioners by providing ready access to new information and ideas from around the world. It can also inhibit reflection because of the temptation to shortcut with a "copy and paste" solution, which may not be the best fit for your situation.

▶ Standards and Codes of Practice can be of great assistance in controlling hazards and ensuring consistent practice across national boundaries; however, reflection is still required to ensure that they continue to be appropriate and commensurate with the risk.

▶ There are a number of tools to aid reflective practice ranging from methods you can use alone, such as reflective learning logs and methods of reflecting with others, one-to-one or in a larger group.

▶ Each reflective tool has its strengths and challenges and they must be understood effectively before employing them.

▶ Reflective learning remains a very personal process in which the individual, his or her conviction of the benefits of the reflection and of course their commitment to using a tool remain the most critical variables.

Notes

1. Hofstede, G.J. and Minkov, M. *Cultures and Organizations: Software of the Mind*, 3rd edn (New York: McGraw-Hill, 2010).
2. Trompenaars, F. and Hampden-Turner, C. *Riding the Waves of Culture: Understanding Diversity in Global Business*, 3rd edn (London: Nicholas Brealey Publishing, 2012).
3. National Fire Protection Association (NFPA) *NFPA 30: Flammable and Combustible Liquids Code (2012)*, USA. Available at: http://www.nfpa.org/codes-and-standards/document-information-pages?mode=code&code=30 (accessed 17 December 2013).
4. Johns, C. *Becoming a Reflective Practitioner*, 2nd edn (Oxford: Blackwell Publishing, 2004).
5. Tuckman, B.W. and Jensen M.A.C. "Stages of Small-Group Development Revisited," *Group and Organization Studies*, 2(4) (1977) (pre-1986). ABI/INFORM Global, p. 419. Available at: http://www.freewebs.com/group-management/BruceTuckman(1).pdf (accessed 2 December 2013).
6. Williams, T. "Learning from Projects", *Journal of the Operational Research Society*, 54 (2003): 443–451.

7. Budworth, T. and Al Hashemi, W.G. "The HSE Practitioner: The Reflective Learner", paper presented at American Society of Safety Engineers – Middle East Chapter, Professional Development Conference, Kingdom of Bahrain, February 2012.

CHAPTER 7

Discussion

Introduction

The development of HSE practitioners both academically and vocationally in health and safety is extremely important. Vocationally, a practitioner can only prove their abilities through demonstrating that they are able to carry out an inspection or audit or a safety case review effectively, identifying the hazards present or latent, evaluating risks and designing control measures which deal with the causes of the problem and are commensurate with the size of the risk.

The underpinning academic knowledge is vital. It helps the practitioner use structured and well-informed approaches; develops the ability to think and analyse more complex situations; challenge accepted approaches and synthesise new solutions.

The evidence that we collected from almost all of the HSE practitioners that we interviewed was that the ability to manage themselves, to influence and achieve through others, to deal with tricky political situations and handle interpersonal issues was the part of the role that they struggled with the most.

There seems to be three areas of development that all HSE practitioners need to focus on and they are:

▶ acquiring, extending and maintaining the underpinning knowledge for competent performance – the subject matter of initial professional training and ongoing professional development;

- ▶ acquiring the technical, practical skills for conducting activities such as risk assessment, accident investigations, audits of management systems, devising policies and procedures;
- ▶ the interpersonal skills to navigate the realities of the role within an organisation, including building relationships, influencing others, and dealing with conflict.

In earlier chapters we reviewed how the HSE practitioner can choose to undertake their initial professional training to develop their knowledge and skills in these areas. Traditionally a professional, in any field, delivers a service – advice or action, or both – based on their specialist knowledge, systematically formulated and applied to the problem of the client. In HSE, as in many other fields, many problems do not neatly fit this model and may involve situations which are "unique, uncertain, and conflicted". The ability to develop and test new forms of understanding and action where familiar categories and ways of thinking fail is central to competent practice as a HSE professional. Developing your professional practice requires "reflection in action", building on your existing understanding by trying different approaches.

A further challenge is for those HSE practitioners who become managers and even directors, working at a more senior, strategic level and responsible for steering their organisation's approach to health, safety, the environment and sustainability. Leading a team and having some responsibility for their development requires a new skill set with underpinning knowledge to be developed. As with initial professional development in HSE, there are a plethora of training opportunities including formal qualifications like postgraduate management diplomas and MBAs, as well as shorter programmes at business schools or provided in-company. Many management development programmes make a deliberate choice to use experiential or reflective learning as a formal part of the course, to encourage the use of reflective tools throughout participants' future working lives.

It is obvious that we all learn from our experiences. However, to gain the maximum from any experience, we must be able to capture it, digest it, analyse and understand it, then extract the learning points and put them into practice in the future. The process is simple but requires a certain conviction, practice, discipline and commitment. In the absence of reflection which is a process of standing back from an experience and taking some time to analyse and carefully review it, some of the learning opportunities may be lost.

The process of reflection can change not just the choices that an individual can make in a situation, but their perception of the situation itself. In critical reflection the change occurs as a person starts to challenge their own firmly held beliefs and thinking.

There was no loss of life but there was a huge loss of property. It was an accident that occurred in the Suez Canal with one of our tankers. It ran aground in the channel of the Suez Canal, and this loss led to the bankruptcy of the company that I was working for.

Up to that moment I could not understand why, for example, why we had to obey some regulations, or do some paperwork or do training or do drills or do whatever. At that time I was actually working on the ships and I had to do all this. For me, it was really a burden and I was uncomfortable because I was thinking that this was only a bureaucratic procedure; that has to be done for the inspectors.

Once this incident occurred, we started from the end trying to find the beginning of the whole story and we saw that many things could have been avoided if we had proper training, proper blah blah blah . . . Then I realised that all these things that I was practising all this time, it was not only for paperwork and bureaucracy, it was because there is a need for these things to be there.

So I got involved . . . on safety issues, trying to prevent incidents happening during mooring operations of ships and I started to see things from quite a different perspective than before.

Because I have been a seaman, I can understand it from the marine position and likewise from the terminal operator's point of view. That's why when we do training, I try to minimise the bureaucratic part of the story, the paperwork. I want them to actually understand what we want. To tell them it doesn't matter if they complete the paperwork perfectly or not; as long as they know what to do and they can comprehend the risks to avoid a problem, any problem, for the terminal. Because if something happens, it is not only the ship that will be affected, it is also the terminal. Terminals mean financial losses and if we lose one terminal for a month or so, then the losses are billions.

Senior Compliance Officer, Marine Safety, Oil and Gas Sector

It is a particular need for those in the HSE profession to learn quickly. HSE is often a second or even third career for many practitioners. This is now changing, but the average experience of HSE practitioners is substantially less than those of a similar age working in more traditional professions such as law, engineering or medicine. To retain credibility, the HSE practitioner needs to devote the time to develop the complex technical, problem-solving, interpersonal and change management skills required to be effective, gain credibility and the respect of those members of other professions that they are working alongside.

Unlike medicine or law, HSE does not have a formal structured training programme for the practitioner to acquire both the underpinning knowledge and the practical skills of their profession. To call yourself a doctor or a barrister, you are required to meet certain strict criteria. Anyone can call themselves a safety advisor, whether they are a member of a Chartered Professional Body or a Board Certified Professional or have no qualifications whatsoever. This situation is partly as a consequence of the relative youth of HSE as a profession. However, ill-qualified people working in the field and the readiness of some organisations to blame unpopular decisions on "health and safety" have led to the profession being held in contempt in some quarters.

Given these challenges, with the positive impact that reflective learning and development can have on the effectiveness of HSE practitioners, it could be argued that reflection should become an essential part of HSE practitioners' initial professional development.

If all roads lead to Rome, then many routes lead to a career in HSE. This can be a real strength in that we bring our experiences and expertise from many different disciplines into the profession. The authors have worked with HSE practitioners who came from such diverse backgrounds as behavioural science, chemistry, HR, nuclear, marine mechanical and civil engineering, as an electrician, quantity surveying and nursing. Whatever our route in, there is the need to acquire a solid foundation of HSE knowledge as part of our initial professional development. No one method of obtaining your initial professional training is inherently better or worse than any other, and the choice depends on resources, opportunity and the preferences of the new practitioner.

Learning has to continue throughout any career as organisational and technological contexts change. This is even more important in a very dynamic profession where hazards, systems, regulations, standards, assessment methods and tools are fast developing. Learning to be reflective can enhance the opportunity to learn "on the job", build one's confidence and competence as a practitioner, and ultimately keeps a practitioner more responsive to emerging situations and challenges.

There are numerous opportunities for development quite apart from those formally provided through training courses. Practitioners must reflect on their needs, their areas of difficulty and critical events that occur and they must then proactively identify opportunities to enhance knowledge and skills. This indeed can be one of the most interesting and rewarding aspects of a career in HSE.

> *In my annual appraisal, there was one particular time that the refinery manager, who was my boss then, said, "You've been with safety for a long time. Don't you get tired?"*
>
> *I said no, because there is always something new to be learned, there is something new to apply, something new out there. All you have to do is be hungry for that information to find it, especially when the internet came in. Google it. It's there; find it, read it, study it, implement it.*
>
> **HSE Advisor, Oil and Gas Sector**

Practitioners are encouraged to consider their colleagues both within and outside of the profession as a rich untapped resource for their own development. Often the skills other professionals develop are of direct benefit to the HSE practitioner. Building your network of mentors both inside the HSE profession and outside maximises your learning opportunities.

Reflective learning does not feel natural to many people, and using reflection as a learning tool for development and self-improvement in a conscious and systematic way is not always practised by many practitioners. A greater understanding of the stages of learning and awareness by the practitioner of his or her own learning style preferences can be very valuable in trying to become more reflective. There are also a variety of tools which can help in the process of becoming a reflective practitioner. It is hoped that this book becomes a valuable resource to HSE practitioners setting out on this journey.

One very useful definition of reflection is the "process of internally examining and exploring an issue of concern triggered by an experience, which creates and clarifies meaning in terms of self and which results in a changed conceptual perspective". In the previous chapters we have discussed how reflection can give you the ability to break out of the context that you are in, to see things in a whole new context, and take a completely different approach to your professional role. This radical change is sometimes triggered by reflection on one critical incident leading to double or triple loop learning.

Reflective or experiential learning and problem-solving processes are closely linked and are summarised in Kolb's learning cycle described and discussed in Chapters 2 and 3. To be effective as reflective practitioners we need to spend time at each stage of the cycle. A greater understanding of the stages of learning and also a better understanding of your own learning style preferences can be extremely empowering and valuable in enhancing your reflective skills. It can allow you to focus on the parts of the cycle that feel less natural to you.

There is a plethora of different types of learning available to HSE practitioners, from your initial training through your entire career. At all stages it is important to give oneself the time to reflect, ideally considering how what you are learning links and connects with or contradicts what you already know and formulating ideas about how you can put it into practice in your work environment. In this way, you maximise the learning opportunity and gain the greatest return on your investment of time and money.

> *A lot of people complain that a conference is too expensive. And actually I've worked for employers that wouldn't stump up the few hundred pounds it cost to go. But I got an idea from a session at a conference which saved my employer £5 million over the course of the next year, so, yes, a very worthwhile return on investment for them.*
>
> **Head of Health and Safety, Utilities Sector**

As we move into management positions, we have an additional responsibility: to encourage and foster a culture of reflection in our teams. That is not just to ensure that we learn lessons from our mistakes – that just tells us what didn't work. It's more important to share the learning from what went really well. As a manager we hold our team members accountable to deliver their key functions and meet their objectives. Learning must form part of these objectives. Action plans with time-bound objectives reviewed at intervals should follow all training courses. Ensuring reflective discussion forms part of team meetings, and requiring team members to share their learning from an event with the rest of the team can develop the habit of reflecting in groups.

As leaders, we set the tone for our team. Our staff know what is important to us; whether we like it or not, we demonstrate it to them with every conversation we have and by the behaviour we exhibit. We need to model good habits ourselves to encourage it in those we lead.

A number of methods for developing our ability to reflect have been discussed in this book. These span a very personal introspective process of self-reflection, from a one-to-one reflective dialogue with a mentor, to a wider group discussion with many individuals focused on experiential learning. Each method has its strengths and weakness. Some may not be available to you at all. Some may feel uncomfortable for you. However, using reflection as a learning tool for development and self-improvement in a conscious and systematic way can help improve the effectiveness of the HSE practitioner, just as it has been used by many other professions over the last few decades. It is worth persisting.

We wrote this book hoping that will it become a valuable resource for HSE practitioners. We wanted to help them in their professional development, as we ourselves were helped by the many, more experienced practitioners and managers we have been privileged to work with. Yet more important than having an interesting and satisfying career, is our calling as HSE professionals, and the duty we owe to those we protect by exercising our professional skills. Our job as a HSE professional is about saving lives and protecting people from the misery of life-changing illness and injury arising from work. We owe it to those we serve to be the very best HSE practitioner that we can be.

Each of the HSE practitioners that we interviewed in the research for this book answered one final question for us. We asked: "If you could go back in time, what advice would you give to your younger self, as you embarked on a career in health and safety or environmental management?" There were many common themes in the answers. This one summarises them all.

> *Looking back fifteen years, I used to think I knew everything in safety . . . Now I see health and safety as an ocean and I'm at one shore. If you think "I know everything", it means you lack understanding.*
>
> *I think you should always try to be attentive and humble. Authoritative, yes, but at the same time, you should be polite in your approach both to management and to the workers. You need to understand their perception of the situation. When it comes to trying to convince them to do something, if your perception is different to theirs, if there is a mismatch, you will achieve nothing.*
>
> *So a good safety officer should have a thorough knowledge of the subject but at the same time just knowledge is not enough. You need to communicate that knowledge in a way that you help people to do the right thing.*
>
> Senior HSE Compliance Officer, Oil and Gas Sector

Appendix

Reflective accounts of writing this book

Teresa

Why did I want to write this book? I asked myself that a number of times over the year that we planned it, conducted the research and worked and re-worked the drafts.

It came about after a conversation that Waddah and I had when we met, probably for the second or third time at a conference in the Middle East. Most of my career has been focused on the development of safety and health practitioners. I must have trained hundreds of practitioners over the 17 years that I spent as a Health and Safety Consultant, guiding many through NEBOSH Qualifications. More recently, I have worked for NEBOSH delivering those qualifications for HSE practitioners through a network of training providers. I have strong views on the value that a good safety practitioner can deliver to an organisation, and, more importantly, the contribution they make to society.

Like many practitioners, my choice of career was inspired by my early experiences. My father worked in the steel industry, suffered one moderately serious accident and then became a safety representative seeking to protect his co-workers. As a teenager, I earned pocket money babysitting – a role that I was well suited to, based on my birth-order as the oldest sister with two younger brothers and more than a hundred fostered siblings joining our family over a period of many years. One of the most devastating experiences of my teens was the death of the father of two young children that I cared for. He fell off a ladder at work. The family lost their father and they lost their home.

It's important we get this right because people's lives depend upon it.

In my early career, I had my qualification – a degree in health and safety – but I was sorely lacking in the practical skills to be a good HSE practitioner, and even more clueless on the social skills. Naïvely I believed that if I gave people technically correct advice, they would follow it.

I am deeply grateful to many of the HSE practitioners, and non-HSE professionals and managers that I was lucky enough to work with and for, who helped me on the way. I wanted to help other HSE practitioners to develop themselves, particularly those who may not be lucky enough to have colleagues as patient and generous as those who mentored me.

One of the aspects of writing this book that I most enjoyed was the opportunity for discussion and debate with HSE practitioners from a wide variety of backgrounds. It was fascinating exploring cultural similarities and differences with HSE practitioners from the UK, Europe, the Middle and Far East and the Indian sub-continent.

As the book took shape, it got more exciting, the swapping of chapters between Waddah and I grew more intense and the "would we/wouldn't we" hit the publisher's deadline became more pressing . . .

One of the most challenging aspects was the impact on my work–life balance. As one of my tutors from Ashridge; the business school I attended for part of this year, asked me: "Why do you spend your leisure time on something which sounds like work?" The answer, of course, is that it is something I believe in. The cost was less time with my family and friends, and often the burden of my chores falling on my husband.

Knowing now how the process has panned out, would I advise my one-year younger self to undertake this project? Yes.

Will I write another one? Hmmm . . . Jury's out.

If I did write another, what are my lessons learned? First, I find an empty page terrifying. Better to bang something down and edit it later rather than trying to get it perfect first time and ending up writing nothing at all. Second, I would schedule the breaks away from the screen, because it made me more productive at the screen. Third, I would empty the house of cheese. It really doesn't help the creative process.

The process of writing of necessity involved a lot of reflection on my own career in safety. Some of this was actually quite painful as I reflected on some of the glaring errors I had made, and situations that I had handled particularly badly. It was quite difficult to stop myself from judging harshly the 24-year-old practitioner that I used to be for falling well short of the standards I would apply to myself now.

I continue to practise reflection in many areas, both within my working life and outside of it. Over the last two or three years I have taken running rather more seriously, and particularly when training for an event, such as a half marathon (I won't call them races because I have no hope of winning), I note my training runs, routes, distances, times and how I felt. One frequent occurrence is that before each event I have gone through a phase where training isn't going well and I believe I'm not going to make it. Actually reflecting back that I felt like this last time helps me to know that I can and will make it round 13.1 miles. Reflective practice keeps me going through the tough times.

When I was younger and more junior, others had little hesitation in pointing out areas for improvement. As you become more senior, perhaps fewer people are willing to share their views on how you can improve. I hope that using reflective practice I can become more like the leader that I would wish to be led by.

Waddah

I only really started to realise the powerful benefits of structured reflective learning in the past few years. Working in a highly demanding job requiring general management, change management and technical development within such a large organisation while studying and working on several really different presentations, papers and books does not allow much time for reflection. This was at least until I made an active and serious effort to take five and take the time to reflect.

I have been fortunate so far to have had an immensely diversified career, rich with interesting and valuable experiences. I have worked with various mentors and some really knowledgeable and experienced people. I became a manager relatively early in my career. This meant I had to learn fast and put leverage on the knowledge of others effectively. I tend to agree with Teresa when she says that the disadvantage of becoming more senior in your job posting is that people may become more and more reluctant to give you frank and candid feedback which you can use to improve. I feel that much of my development has come through reflection, though it has only become more structured in the past few years. I also currently have a boss who is a "people's man" who is greatly empathetic and has over the years helped me address organisational challenges by sharing deeply reflective accounts from his career.

I must say that I feel I have learnt as much from reading and researching in the past three years of my DBA study as I have from interacting with others at work, university and in my personal life. In fact, I am glad that part of our DBA module requirements was to prepare a reflective journal over two years. I say I am glad because if I had not been forced to read and explore reflective learning in a structured way, I might not have learned and gained so much from it.

I started to feel that (and I think it is just a sign of getting older), having greater breadth of experience and being also in a position at times to advise others, that a reflective approach was one that others could benefit from. I know when Teresa and I spoke initially about putting a technical conference paper together on reflective learning and HSE practitioners, it was going to be an important contribution to our community and I really hope it was.

Through our peer reviews of the chapters of this book and the discussions on the subject with Teresa, I actually started to appreciate even more how important reflective learning really was to a HSE practitioner. My other interests lie in theological philosophy and while it may seem that this is so remotely connected to my profession and academic training and education, when I reflect on matters of deep purpose, causality, connectedness and rationality, I find my knowledge, skills and experiences all start to connect.

In the past few years I have personally gained so much from the true love of my life my wife with whom I share a great reflective relationship. As my kids have been growing, reflecting has become even more important as one finds oneself having to slow down and speed up your thinking process to communicate with them more effectively and keep up with their incredibly rational questioning respectively.

I remember in the past year and when she was still not even 5 years old, my daughter asking me, "Father, this is your office, right?" (she was talking to me in my study at home where I do most of my reading and writing), and I said, "Yes, dear"– then she asked me, "So, you do your studying here?" and I said, "Yes, dear" – and then she asked me if I also worked (on my job-related work) in my study and I said, "Yes, dear" – so she said, "So why do you need to drive every morning to your office in town to work when you can do your work here at home, and then you can also cook lunch for us every day and play with us as soon as we are back from school?" I reflected very long on this conversation, as though it may not be a practical suggestion – in terms of pure rationality, my child was very right as this was borne from her line of questioning which was uninhibited by pre-set assumptions.

I feel that we limit ourselves by pre-set assumptions based on acquired knowledge and while this is how we develop, it can also be how we inhibit our development. I have and continue to reflect on many aspects in my working and personal life. When we were working on this book, like Teresa, I kept asking myself questions: have we covered all the ground? Is it too academic? Is it too serious? Are there enough examples? But most importantly, will it motivate my fellow practitioners to engage with reflective learning?

I have conducted a couple of workshops over the past two years on this subject and I have found that the concept of reflective and experiential learning is generally attractive to HSE practitioners. I have also found great differences in how practitioners were willing to engage with their own reflective and experiential learning.

Lastly, I think the foundation of my reflective practice was definitely nurtured by both my wonderful parents who have always encouraged me to think, explore, question and reflect. In my current research with my father on the seven most oft-repeated verses of the Holy Quran, "*Surat Al Fatiha*", I would say that one of the most enjoyable aspects of the five-year project has been the reflective dialogues we continue to have.

It may seem strange for one to say this, having written this book, but I will definitely continue to read it back to myself from time to time, as I feel that as I grow older and mature further as a practitioner, my affinity for certain reflective tools may change!

Bibliography

Anders Ericsson, K., Krampe, R.T. and Tesch-Romer, C. (1993) "The Role of Deliberate Practice in the Acquisition of Expert Performance", *Psychological Review*, 100(3): 363–406.

Argyris, C. and Schön, D. (1996)

Beard, C. and Wilson, J.P. (2006) *Experiential Learning: Best Practice Handbook for Educators and Trainers*, 2nd edn, London: Kogan Page.

Bloom, B.S. (ed.) (1956) *Taxonomy of Educational Objectives: The Classification of Educational Goals: Handbook I: Cognitive Domain*, New York: Longman.

Bostrom, R.P., Olfman, L. and Sein, M.K. (1990) "The Importance of Learning Style in End-User Training", *MIS Quarterly*, 14(1): 101–110.

Boud, D., Keogh, R. and Walker, D. ([1985] 2005) *Reflection: Turning Experience into Learning*, London: RoutledgeFalmer.

Budworth, T. and Al Hashemi, W.G. (2012) "The HSE Practitioner: The Reflective Learner", paper presented at American Society of Safety Engineers, Middle East Chapter, Professional Development Conference, Kingdom of Bahrain, February.

Chivers, G. (2003) "Utilizing Reflective Practice Interviews in Professional Development", *Journal of European Industrial Training*, 27(1): 5–15.

Cialdini, R. (1993, 2007) *Influence: The Psychology of Persuasion*, New York: HarperBusiness, rev. edn, 1 Feb. 2007.

Doyle, W. and Young, J.D. (2000) "Management Development: Making the Most Out of Experience and Reflection", *The Canadian Manager*, 25(3).

ENSHPO Certification Standard for European Safety and Health Managers (EUROSHM) (2013). Available at: http://www.euroshm.org/full.php (accessed 15 December 2013).

Eraut, M. and Steadman, S. (1998) "Evaluation of Level 5 Management S/NVQs Final Report 1998", Research Report Number 7, Brighton: University of Sussex.

Garvin, D. (2000) *Learning in Action: A Guide to Putting the Learning Organization to Work*, Boston: Harvard Business School Press.

Gladwell, M. (2009) *Outliers: The Story of Success*, Harmondsworth: Penguin.

Gray, D. (2007) "Facilitating Management Learning: Developing Critical Reflection through Reflective Tools", *Management Learning*, 28(5): 495–517.

Health and Safety Executive (1997) *Successful Health and Safety Management* HSG65, London: HSE, 2nd edn. Available at: free-to-download, http://www.hseni.gov.uk/hsg65_successful_h_s_management.pdf.

Hey, A., Peltier, J.W. and Drago, W.A. (2004) "Reflective Learning and On-Line Management Education: A Comparison of Traditional and On-Line MBA Students", *Strategic Change*, 13(4): 169–180.

Hofstede, G.J. and Minkov, M. (2010) *Cultures and Organizations: Software of the Mind*, 3rd edn, New York: McGraw-Hill.

Honey, P. (1983) "Building on Learning Styles", *Training Officer*, April, Action IMC University Press. Reference LTL 141/147. Available at: http://www.peterhoney.com/ (accessed 14 October 2013).

Honey, P. and Mumford, A. (1982) *Manual of Learning Styles*, London: Peter Honey Publications.

Hughes, E.C. (1963) "The Professions", *Daedalus*, 92(4): 655–668.

Illeris, K. (2004) *Three Dimensions of Learning*, Malabar, FL: Krieger Publishing.

Jensen, P.E. (2005) "A Contextual Theory of Learning and the Learning Organization", *Knowledge and Process Management*, 12(1): 53–64.

Johns, C. (2004) *Becoming a Reflective Practitioner*, 2nd edn, Oxford: Blackwell Publishing.

Kelly, G.A. (1955) *The Psychology of Personal Constructs*, New York: W.W. Norton.

Kolb, D. (1984) *Experiential Learning: Experience as the Source of Learning and Development*, Englewood Cliffs, NJ: Prentice Hall.

MacFarlane, B. (2001) "Developing Reflective Students: Evaluating the Benefits of Learning within a Business Ethics Programme", *Teaching Business Ethics*, 5(4): 375–387.

Marcus, G. (2012) *Guitar Zero: The Science of Becoming Musical at Any Age*, Harmondsworth: Penguin.

Moon, J. (1999) *Learning Journals: A Handbook for Academic Students and Professional Development*, London: RoutledgeFalmer.

Moon, J. (2009) *A Handbook of Reflective and Experiential Learning: Theory and Practice*, London: RoutledgeFalmer.

National Examination Board in Occupational Safety and Health (2008) *NEBOSH Guide to the Diploma in Occupational Health and Safety*, Leicester: NEBOSH.

National Fire Protection Association (NFPA) (2012) *NFPA 30: Flammable and Combustible Liquids Code (2012) USA*. Available at: http://www.nfpa.org/codes-and-standards/document-information-pages?mode=code&code=30 (accessed 17 December 2013).

Office of Qualifications and Examinations Regulation (OfQual) (n.d.) "Regulated Qualifications Activity Dataset, 2007/8 to Present". Available at: http://ofqual.gov.uk/standards/statistics/raw-data (accessed 15 December 2013).

Ormrod, J.E. (1995) *Human Learning*, Englewood Cliffs, NJ: Prentice Hall.

Peters, K. (2011) "Neuroscience, Learning and Change 360°", *The Ashridge Journal: Perspectives*, Spring.

Raelin, J.A. (2001) "Public Reflection as the Basis of Learning", *Management Learning*, 32(1): 11–30.

Rasmussen, J. (1982) "Human Errors: A Taxonomy for Describing Human Malfunction in Industrial Installations", *Journal of Occupational Accidents*, 4: 311–333.

Reason, J. (1990) *Human Error*, Cambridge: Cambridge University Press.

Schein, E. (1974) *Professional Education*, New York: McGraw-Hill.

Schön, D.A. (1987) *Educating the Reflective Practitioner: Toward a New Design for Teaching and Learning in the Professions*, San Francisco: Jossey-Bass Publishers.

Scott, J.L. (2002) "Awareness of Actual Learning Processes", *Journal of the Operational Research Society*, 53(1): 2–10.

Sims, R.R. and Sims, S.J. (1995) *The Importance of Learning Styles*, Westport, CT: Greenwood Publishing Group, Inc.

Smith, B. (1993) "Building Managers from the Inside Out: Developing Managers through Competency-Based Action Learning", *Journal of Management Development*, 12(1): 43–48.

Smith, M.K. (2001) "Chris Argyris: Theories of Action, Double-Loop Learning and Organizational Learning", in *The Encyclopaedia of Informal Education*. Available at: http://infed.org/mobi/chris-argyris-theories-of-action-double-loop-learning-and-organizational-learning/ (accessed 8 December 2013).

Syed, M.B. (2011) *Bounce: The Myth of Talent and the Power of Practice*, London: Fourth Estate.

Trompenaars, F. and Hampden-Turner, C. (2012) *Riding the Waves of Culture: Understanding Diversity in Global Business*, 3rd edn, London: Nicholas Brealey Publishing.

Tuckman, B.W. and Jensen, M.A.C. (1977) "Stages of Small-Group Development Revisited", *Group and Organization Studies*, 2(4). Available at: http://www.freewebs.com/group-management/BruceTuckman(1).pdf (accessed 2 December 2013).

Williams, T. (2003) "Learning from Projects", *Journal of the Operational Research Society*, 54: 443–451.

Index